2nd International Workshop on New Computational Methods for Inverse Problems

AA008653

(NCMIP 2012)

Journal of Physics: Conference Series Volume 386

Cachan, France
15 May 2012

ISBN: 978-1-62276-347-4
ISSN: 1742-6588

Printed from e-media with permission by:

Curran Associates, Inc.
57 Morehouse Lane
Red Hook, NY 12571

Some format issues inherent in the e-media version may also appear in this print version.

Printed by Curran Associates, Inc. (2013)

For permission requests, please contact the Institute of Physics
at the address below.

Institute of Physics
Dirac House, Temple Back
Bristol BS1 6BE UK

Phone: 44 1 17 929 7481
Fax: 44 1 17 920 0979

techtracking@iop.org

Additional copies of this publication are available from:

Curran Associates, Inc.
57 Morehouse Lane
Red Hook, NY 12571 USA
Phone: 845-758-0400
Fax: 845-758-2634
Email: curran@proceedings.com
Web: www.proceedings.com

TABLE OF CONTENTS

This volume of *Journal of Physics: Conference Series* is dedicated to the scientific contributions presented during the 2nd International Workshop on New Computational Methods for Inverse Problems, (NCMIP 2012). This workshop took place at Ecole Normale Supérieure de Cachan, in Cachan, France, on 15 May 2012, at the initiative of Institut Farman. The first edition of NCMIP also took place in Cachan, France, within the scope of the ValueTools Conference, in May 2011 (http://www.ncmip.org/2011/).

The NCMIP Workshop focused on recent advances in the resolution of inverse problems. Indeed inverse problems appear in numerous scientific areas such as geophysics, biological and medical imaging, material and structure characterization, electrical, mechanical and civil engineering, and finance. The resolution of inverse problems consists of estimating the parameters of the observed system or structure from data collected by an instrumental sensing or imaging device. Its success firstly requires the collection of relevant observation data. It also requires accurate models describing the physical interactions between the instrumental device and the observed system, as well as the intrinsic properties of the solution itself. Finally, it requires the design of robust, accurate and efficient inversion algorithms. Advanced sensor arrays and imaging devices provide high rate and high volume data; in this context, the efficient resolution of the inverse problem requires the joint development of new models and inversion methods, taking computational and implementation aspects into account. During this one-day workshop, researchers had the opportunity to bring to light and share new techniques and results in the field of inverse problems.

The topics of the workshop were: algorithms and computational aspects of inversion, Bayesian estimation, kernel methods, learning methods, convex optimization, free discontinuity problems, metamodels, proper orthogonal decomposition, reduced models for the inversion, non-linear inverse scattering, image reconstruction and restoration, applications (bio-medical imaging, non-destructive evaluation etc).

NCMIP 2012 was a one-day workshop. Each of the submitted papers was reviewed by 2 to 4 reviewers. Among the accepted papers, there are 8 oral presentations and 5 posters. Three international speakers were invited for a long talk. This second edition attracted 60 registered attendees in May 2012.

NCMIP 2012 was supported by Institut Farman (ENS Cachan) and endorsed by the following French research networks (GDR ISIS, GDR Ondes, GDR MOA, GDR MSPC). The program committee acknowledges the following laboratories CMLA, LMT, LSV, LURPA, SATIE, as well as DIGITEO Network.

Laure Blanc-Féraud and Pierre-Yves Joubert

Workshop Co-chairs
Laure Blanc-Féraud, I3S laboratory, CNRS, France
Pierre-Yves Joubert, IEF laboratory, Paris-Sud University, CNRS, France

Technical Program Committee
Alexandre Baussard, ENSTA Bretagne, Lab-STICC, France
Marc Bonnet, ENSTA, ParisTech, France
Jerôme Darbon, CMLA, ENS Cachan, CNRS, France
Oliver Dorn, School of Mathematics, University of Manchester, UK
Mário Figueiredo, Instituto Superior Técnico, Lisbon, Portugal
Laurent Fribourg, LSV, ENS Cachan, CNRS, France
Marc Lambert, L2S Laboratory, CNRS, SupElec, Paris-Sud University, France
Anthony Quinn, Trinity College, Dublin, Ireland
Christian Rey, LMT, ENS Cachan, CNRS, France
Joachim Weickert, Saarland University, Germany

Local Chair
Alejandro Mottini, Morpheme group I3S-INRIA
Sophie Abriet, SATIE, ENS Cachan, CNRS, France
Béatrice Bacquet, SATIE, ENS Cachan, CNRS, France

Reviewers
Gilles Aubert, J-A Dieudonné Laboratory, CNRS and University of Nice-Sophia Antipolis, France
Alexandre Baussard, ENSTA Bretagne, Lab-STICC, France
Laure Blanc-Féraud, I3S laboratory, CNRS, France
Marc Bonnet, ENSTA, ParisTech, France
Jerôme Darbon, CMLA, ENS Cachan, CNRS, France
Oliver Dorn, School of Mathematics, University of Manchester, UK
Gérard Favier, I3S laboratory, CNRS, France
Mário Figueiredo, Instituto Superior Técnico, Lisbon, Portugal
Laurent Fribourg, LSV, ENS Cachan, CNRS, France
Jérôme Idier, IRCCyN, CNRS, France
Pierre-Yves Joubert, IEF laboratory, Paris-Sud University, CNRS, France
Marc Lambert, L2S Laboratory, CNRS, SupElec, Paris-Sud University, France
Dominique Lesselier, L2S Laboratory, CNRS, SupElec, Paris-Sud University, France
Anthony Quinn, Trinity College, Dublin, Ireland
Christian Rey, LMT, ENS Cachan, CNRS, France
Simon Setzer, Saarland University, Germany
Joachim Weickert, Saarland University, Germany

Invited speakers
Antonin Chambolle: CMAP, Ecole Polytechnique, CNRS, France

Matteo Pastorino: University of Genoa, Italy
Michael Unser: Ecole polytechnique Fédérale de Lausanne, Switzerland

Journal of Physics: Conference Series **386** (2012) 012001 doi:10.1088/1742-6596/386/1/012001

A formalization of the *flutter shutter*

Yohann Tendero, Bernard Rougé, Jean-Michel Morel

Centre de Mathématiques et de Leurs Applications (CMLA)
École Normale Supérieure de Cachan - 61 av du Pdt Wilson 94235 Cachan Cedex France.

E-mail: tendero@cmla.ens-cachan.fr

Abstract. Acquiring good quality images of moving objects by a digital camera remains a valid question. If the velocity of the photographed object is not known, it is virtually impossible to tune an optimal exposure time. For this reason the recent Agrawal *et al.* *flutter shutter* apparatus has generated much interest. In this communication, we propose a mathematical formalization of a general *flutter shutter* method, also permitting non-binary shutter sequences. Thanks to this formalization, the question of the optimal *flutter shutter* code can be defined and solved. The method gives analytic formulas for the best attainable SNR for the restored image. It also gives a way to compute optimal flutter shutter codes.

1. Introduction

If a scene being photographed moves during the exposition process, or if the scene is steady and the camera moves, the resulting images can be degraded by motion blur. Thus, moving scenes require a "small" exposure time, and the number of sensed photons remains limited. The difficulty of motion blur is illustrated by its simplest example, the one dimensional uniform motion blur. The result is a convolution of the image with a one dimensional window shaped kernel. The support of the kernel increases linearly with the exposure time and the velocity of the motion. As soon as the exposure time exceeds a limit, the blur is no more invertible [1]. The restoration process becomes an ill posed problem. An exciting alternative to this classic photography dilemma was proposed in [2] where the authors suggest modifications in the acquisition process to get invertible blur kernels by using a *flutter shutter*. If the shutter sequence is well chosen, invertibility is guaranteed for blurs with arbitrary size supports (see Figs 1, 2). Replacing the classic camera shutter by a *flutter shutter*, it becomes possible to use any integration time. This also means that the exposure time on a given scene can become much longer: many more photons are therefore sensed. Thus, at first sight, the *flutter shutter* looks like a magic solution that should equip all cameras. The question arises of whether it can indefinitely increase the SNR by increased exposure. As we shall see, the answer is that the *flutter shutter* does not permit an indefinite increase of the SNR, no matter how long the exposure time is. This fact, that we will call the "*flutter shutter paradox*", will be translated into closed formulae for the optimal *flutter shutter* and for the SNR of the deconvolved image. The theory will be complemented with simulated experiments that can be also executed on line at Image Processing On Line [3].

In section 2, we reformulate carefully the first steps of the image acquisition model using a physical Poisson model for the photons capture process. The *flutter shutter* is formalized in section 3, where the SNR estimates are also given. This communication ends with an unexpected mathematical answer to the question of the optimal *flutter shutter* code. (*Flutter shutter* codes have usually been optimized by random searches [4, 5, 6, 7, 8]). The proposed formalism and results also apply to the recent *motion-invariant photography* [9], where the authors suggest to accelerate the camera during the exposure, to get an invertible blur.

2. Image acquisition model

Formalizing the *flutter shutter* requires an accurate continuous stochastic model of the photon capture by a sensor array. Without loss of generality the formalization will be done in the case where the sensor array is 1D and where the photographed object is conceived as a "landscape" moving in a direction parallel to the sensor array. Let $\mathbf{P}_l : \mathbb{R}^+ \times \mathbb{R}$ be a bi-dimensional Poisson process of intensity $l(t,x)$, $\forall (t,x) \in \mathbb{R}^+ \times \mathbb{R}$ (here l is called landscape, t resp. x are the time resp. spatial positions). In the classic still photography setup, $l(t,x) = l(x)$, thus the observation of a pixel sensor (photon counter) of unit length centered at x using a exposure time of Δt is a Poisson random variable $\mathbf{P}_l\left([0,\Delta t] \times [x - \frac{1}{2}, x + \frac{1}{2}]\right) \sim \mathcal{P}\left(\int_0^{\Delta t}\int_{x-\frac{1}{2}}^{x+\frac{1}{2}} l(y)dydt\right) \sim \mathcal{P}\left(\Delta t(\mathbb{1}_{[-\frac{1}{2},\frac{1}{2}]} * l)(x)\right)$
where Δt is the exposure time, $[x - \frac{1}{2}, x + \frac{1}{2}]$ represents the normalized sensor unit, $X \sim P$ means that a random variable X has law P, and $*$ denotes the convolution ($(f * g)(x) := \int_{-\infty}^{+\infty} f(y)g(x-y)dy$). For sampling purposes we assume that the theoretical landscape l is seen through an optical system with a point spread function g.

Definition We call ideal landscape the deterministic function $u = \mathbb{1}_{[-\frac{1}{2},\frac{1}{2}]} * g * l$, where g is the *point spread function* (PSF) of the optical system providing a cutoff frequency.

In other words, thanks to the convolution with g the acquisition system is able to sample u. We shall denote by $u(x)$ the ideal (noiseless) pixel landscape value at a pixel centered at x, as it could only be obtained after infinite exposure. Notice that the landscape u contains in itself all spatial integrations required, from the PSF g and from the normalized pixel sensor. We assume in the sequel that $u \in L^1 \cap L^2(\mathbb{R})$.

2.1. Sampling, interpolation

Since the optical kernel g provides a cutoff frequency, u is band-limited, namely $\hat{u}(\xi)(:= \int_{-\infty}^{\infty} u(x)e^{-ix\xi}dx$) is supported in $[-\pi, \pi]$. It could therefore be sampled at unit rate. The discrete sensor observations, or samples, will be denoted by $e(n)$ for $n \in \mathbb{Z}$.

Given a discrete array observation $e(n)$, $n \in \mathbb{Z}$, its band limited interpolate $e(x)$ $x \in \mathbb{R}$ is defined by the Shannon-Whittaker interpolation as

$$e(x) = \sum_{n \in \mathbb{Z}} e(n) sinc(x - n) \qquad (sinc(x) = \frac{sin(\pi x)}{\pi x}) \tag{1}$$

2.2. Noise measurement

We call signal to noise ratio (SNR) of a random variable X the ratio $SNR(X) := \frac{|\mathbb{E}X|}{\sqrt{var(X)}}$.
For example if $\hat{\mathbb{u}}_{est}(\xi)$ is an estimation of the landscape $\hat{u}(\xi)$ based on a noisy observation of u, we call "spectral SNR" of $\mathbb{u}_{est}$ the frequency dependent ratio defined by

$$SNR^{spectral}(\hat{\mathbb{u}}_{est}(\xi)) := \frac{|\mathbb{E}\hat{\mathbb{u}}_{est}(\xi)|}{\sqrt{(var(\hat{\mathbb{u}}_{est}(\xi))}} \qquad \text{for } \xi \in [-\pi, \pi]; \tag{2}$$

Likewise, we call "spectral-averaged SNR" :

$$SNR^{averaged}(\hat{\mathbb{u}}_{est}) := \frac{\frac{1}{2\pi}\int |\mathbb{E}\hat{\mathbb{u}}_{est}(\xi)|\mathbb{1}_{[-\pi,\pi]}(\xi)d\xi}{\sqrt{\frac{1}{2\pi}\int var(\hat{\mathbb{u}}_{est}(\xi)\mathbb{1}_{[-\pi,\pi]}(\xi))d\xi}}. \tag{3}$$

From now on we assume $l(t,x) = l(x - tv(t))$, and mainly $v(t) \equiv v$. Hence all the former discussion made on the acquisition system, sampling and interpolation holds.

3. The *numerical flutter shutter*

The *numerical flutter shutter* method consists in gain modifications taking place *after* the acquisition by the sensor. This numerical set up of the *flutter shutter* is possible for example with CMOS sensors, or with a CCD matrix with adequately chosen gains at each sensor. The set up can also be described as a camera taking a burst of N images with an exposure time Δt. The k-th image is multiplied, for $k \in 0, ..., N-1$, by an $\alpha_k \in \mathbb{R}$ gain. Then all images are added up to obtain *one* observed image. The *analog flutter shutter* considered in [2] is a bit more restrictive: indeed the use of a physical shutter in front of the lens is only possible with positive gains. For a sake of conciseness in this communication, we will only state our results for the *numerical flutter shutter*, but homologue statements exist for the *analog flutter shutter*. An obvious objection to the utility of the *numerical flutter shutter* is that one could keep all images, or perform a multi-image denoising instead of adding them all. The set up is nevertheless relevant for Earth observation satellites, where an uncontrolled blur is caused by a drift in TDI's (time differential integrators). In that case the resulting blur estimate could be eliminated by a *flutter shutter*, without any additional transmission burden if only the *flutter shutter* image (the sum) was transmitted. The k-th acquired elementary image at a pixel at position n is a realization of $\mathcal{P}\left(\int_{k\Delta t}^{(k+1)\Delta t} u(n - vt)dt\right)$. The *flutter shutter* observation is obtained by combining the k-th output with weight α_k. Thus the *flutter shutter* output at a pixel centered at n is

$$obs(n) \sim \sum_{k=0}^{N-1} \alpha_k \mathcal{P}\left(\int_{k\Delta t}^{(k+1)\Delta t} u(n - vt)dt\right) \tag{4}$$

where by construction $obs(n)$ are obtained for $n \in \mathbb{Z}$ and are independent. In the following it will be useful to associate with the *flutter shutter* its *code* defined as the vector $(\alpha_k)_{k=0,\cdots,N-1}$, and its *flutter shutter function* defined by $\alpha(t) = \alpha_k$ for $t \in [k, k+1[$.

Definition Let $(\alpha_0, ..., \alpha_{N-1}) \in \mathbb{R}^N$ be a *flutter shutter* code. We call *flutter shutter* samples at position n of the landscape u at velocity v the random variable

$$obs(n) \sim \sum_{k=0}^{N-1} \alpha_k \mathcal{P}\left(\int_{k\Delta t}^{(k+1)\Delta t} u(n - vt)dt\right). \tag{5}$$

We call *flutter shutter* its band limited interpolate $obs(x) \sim \sum_{n\in\mathbb{Z}} obs(n)sinc(x - n)$. We call *flutter shutter function* the function $\alpha(t) = \sum_{k=0}^{N-1} \alpha_k \mathbb{1}_{[k,k+1[}(t)$.

Theorem 3.1. *The observed samples of the flutter shutter are such that, for* $n \in \mathbb{Z}$

$$\mathbb{E}\left(obs(n)\right) = \left(\frac{1}{v}\alpha(\frac{\cdot}{v\Delta t}) * u\right)(n) \text{ and } var(obs(n)) = \left(\frac{1}{v}\alpha^2(\frac{\cdot}{v\Delta t}) * u\right)(n). \tag{6}$$

Figure 1. Left: simulated noisy and blurry image acquired by a *flutter shutter* camera with a 52 pixels blur support. Right: restored image. Those images were generated from a flutter shutter camera simulator, `http://dev.ipol.im/~tendero/ipol_demo/flutter_demo/`. It simulates a flutter shutter camera assuming a Poisson photon emission. *To deconvolve: 1) apply a classic mirror symmetry in the x-direction 2) compute the DFT and divide it by $\hat{\alpha}$ 3) compute the DFT inverse and crop to remove the mirror symmetry.*

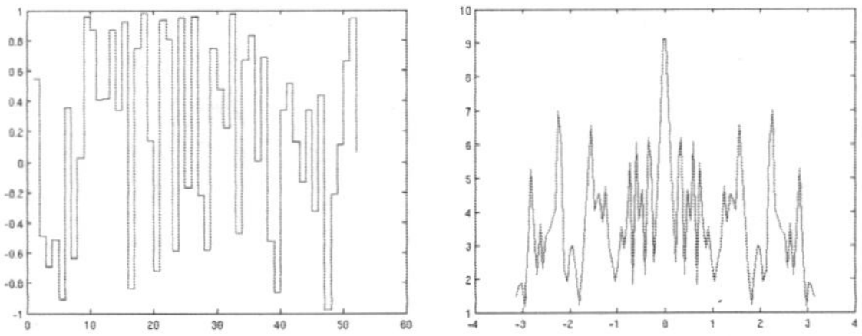

Figure 2. On the left a *flutter shutter* function α; x-axis k, y-axis: gain. On the right its Fourier transform; x-axis ξ, y-axis $|\hat{\alpha}(\xi)|$. Notice that $\hat{\alpha}(\xi) \neq 0 \ \forall \xi \in [-\pi, \pi]$.

3.1. The inverse filter of a flutter shutter

Let us examine first the discrete noiseless case, when $obs(n) = \left(\frac{1}{v}\alpha(\frac{\cdot}{v\Delta t}) * u\right)(n)$ and $obs(n)$ is obtained for $n \in \mathbb{Z}$ but being band limited, can be interpolated to $obs(x)$, for any $x \in \mathbb{R}$. Then $\mathcal{F}\left(\frac{1}{v}\alpha(\frac{\cdot}{v\Delta t}) * u\right)(\xi) = \Delta t \hat{u}(\xi)\hat{\alpha}(\xi v \Delta t)$. We assumed that $\hat{u}(\xi) = 0$ for $|\xi| > \pi$. Hence for the invertibility we must only require that $|\hat{\alpha}(\xi v \Delta t)| > 0$ for $\xi \in [-\pi, \pi]$.

Definition We say that the *flutter shutter* with code $(\alpha_k)_{k=0,\cdots,N-1}$ is invertible if $|\hat{\alpha}(\xi)| > 0$ for $\xi \in [-\pi v_0 \Delta t, \pi v_0 \Delta t]$.

If the *flutter shutter* is invertible, we can consider the inverse filter γ defined by $\hat{\gamma}(\xi) = \frac{\mathbb{1}_{[-\pi,\pi]}(\xi)}{\Delta t \hat{\alpha}(\xi v \Delta t)}$. We shall as usual define the recovered landscape from noisy data by the formulae that would be valid for noiseless data. Assume we observe $e(n) = \mathbb{E}(obs(n))$ for $n \in \mathbb{Z}$ and wish to compute $\hat{e}(\xi)$ from the discrete observed $(e(n))_{n\in\mathbb{Z}}$. Since $e(x)$ is band limited, we can interpolate it using the band limited interpolation (1). The band limited interpolate of the ideal observation is

$$e(x) = \sum_{n\in\mathbb{Z}} e(n) sinc(x - n). \tag{7}$$

Then from (7) we have $\hat{e}(\xi) = \sum_{n\in\mathbb{Z}} e(n)e^{-in\xi}\mathbb{1}_{[-\pi,\pi]}(\xi)$. So the ideal deconvolved landscape $d(x)$ obtained by combining the previous and γ is $\hat{d}(\xi) = \frac{\sum_{n\in\mathbb{Z}} e(n)e^{-in\xi}\mathbb{1}_{[-\pi,\pi]}(\xi)}{\Delta t \hat{\alpha}(\xi v \Delta t)}$. We shall now adopt the same formulae for the noisy case.

Definition Assume that the *flutter shutter* with code $(\alpha_k)_{k=0,\cdots,N-1}$ is invertible. We call estimated landscape $\mathfrak{u}_{est,num}$ of the *flutter shutter* the function defined by (using the obtained $obs(n)$ samples (4) instead of the ideal $e(n)$ in $\hat{d}(\xi)$) $\mathcal{F}(\mathfrak{u}_{est,num})(\xi) = \frac{\sum_{n\in\mathbb{Z}} obs(n)e^{-in\xi}\mathbb{1}_{[-\pi,\pi]}(\xi)}{\Delta t \hat{\alpha}(\xi v \Delta t)}$.

Theorem 3.2. *Given any flutter shutter function α, the flutter shutter has a spectral SNR (2) equal to* $SNR(\xi) = \mathbb{1}_{[-\pi,\pi]}(\xi)\frac{\sqrt{\Delta t}|\hat{u}(\xi)||\hat{\alpha}(\xi v \Delta t)|}{\sqrt{||u||_{L^1}}||\alpha||_{L^2}}$.

Theorem 3.3. *Given a landscape $u(x-vt)$ moving at velocity v, the ideal flutter shutter function is* $\alpha^*(t) = sinc(tv\Delta t)$.

This means that the best $SNR^{averaged}$ of the recovered image is achieved by the *sinc* function. Notice that this *flutter shutter function* has a constant Fourier transform on the support of $\hat{u}$ for any velocity $|\tilde{v}| \leq |v|$. This means that this *flutter shutter* code is (Figs. 3, 4) "self-deconvolving". Being non piecewise-constant (coming from a code) this function is *not directly* implementable using a *numerical flutter shutter* strategy but it can be proved that a piecewise constant equivalent function can be used instead.

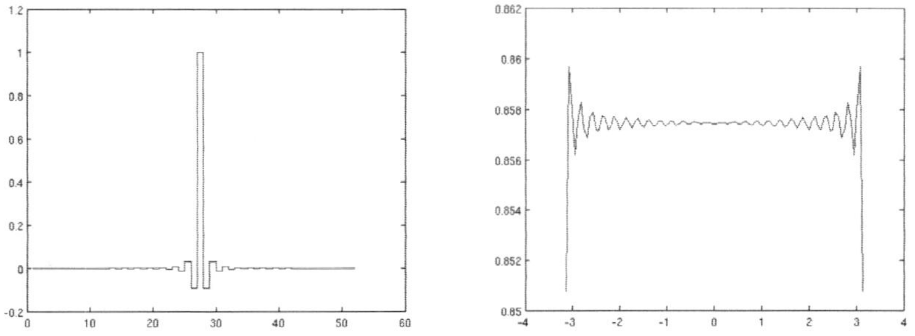

Figure 3. The *flutter shutter* gain function for the sinc-code (left). The Fourier transform (modulus) of the *sinc − code* (right), approximating the Fourier transform of the ideal gain function.

Corollary 3.4. *The ideal flutter shutter has an averaged SNR (3) equal to* $SNR^{averaged} = \frac{1}{2\pi\sqrt{v}} \frac{\int_{-\pi}^{\pi} |\hat{u}(\xi)| d\xi}{\sqrt{\|u\|_{L^1}}}$.

4. Conclusion

These results mean that the SNR of the *flutter shutter* remains bounded despite the use of an infinite exposure time accumulating many more photons than a short snapshot. Given a

Figure 4. *Sinc* code : observed image (left). The blur interval length is equal to 52 pixels here. Reconstructed image ($RMSE$ = 1.33) (middle). Residual noise (difference between ground truth and reconstructed, dynamic normalized on $[0, 255]$ by an affine contrast change). The acquired image is "sharp", it is no surprise since the *sinc*-code has a nearly constant Fourier transform thus, it does not alter any frequency.

reference velocity v for the observed object, it can also be proved *analytically* that among snapshots (classic shutter functions of the form $\alpha(t) = \mathbb{1}_{[0,\Delta t]}(t)$), the best choice is $\Delta t^* \approx \frac{1.0707}{|v|}$ (see Fig. 5). The ratio of $SNRs$ between the best *flutter shutter* and the best snapshot is ≈ 1.17. In other words the *flutter shutter* provides a slight SNR improvement, even when the motion is known. A paradoxical conclusion is this: for the agile numerical cameras of the future, the best camera exposure strategy will always be a *flutter shutter*.

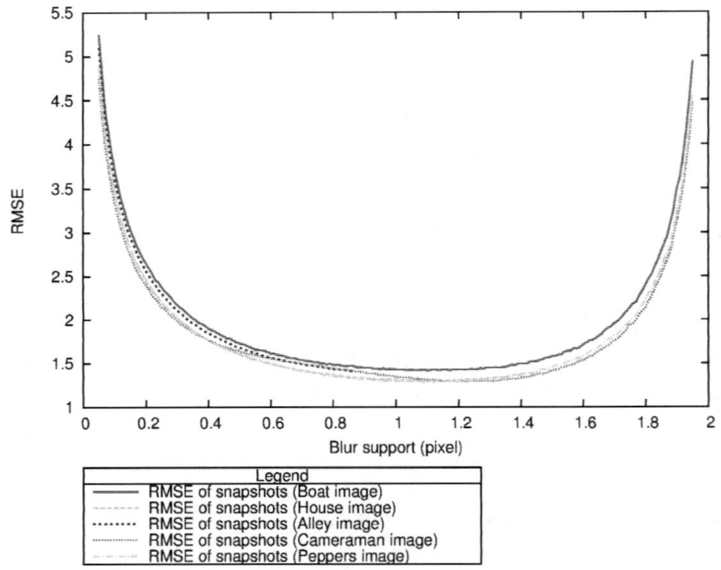

Figure 5. This figure shows the $RMSE$ curves for different snapshots kinds, on five test images (House, Alley, Boat, Cameraman, Peppers). On the $x-axis$, the blur support ($|v|\Delta t$) in pixels, on the $y-axis$ the corresponding $RMSE$. The curves confirm that, on average, the blur support for a standard camera should be of approximatively $\Delta t^* = 1.0909$ pixels. A larger support would lead to a better SNR on the observed image samples, but the deconvolution would entail a lower SNR on the deconvolved image. The best snapshot is a compromise between the number of photons caught during a time span Δt and the deconvolution kernel. It gives a reference to compare all *flutter shutter* strategies in terms of SNR.

References

[1] Boracchi G and Foi A 2011 *Image Processing, IEEE Transactions on* **20** 592–598
[2] Agrawal A and Raskar R 2007 *Computer Vision and Pattern Recognition (CVPR), 2007 IEEE Conference on* (IEEE) pp 1–8
[3] Tendero Y The flutter shutter camera simulator `http://dev.ipol.im/~tendero/ipol_demo/flutter_demo/`
[4] Agrawal A and Xu Y 2009 *Computer Vision and Pattern Recognition (CVPR), 2009 IEEE Conference on* (IEEE) pp 2066–2073
[5] Jelinek J 2010 *Society of Photo-Optical Instrumentation Engineers (SPIE) Conference Series* vol 7701 p 18
[6] Raskar R, Tumblin J and Agrawal A 2009 Method for deblurring images using optimized temporal coding patterns US Patent 7,580,620
[7] Raskar R 2010 Method and apparatus for deblurring images US Patent 7,756,407
[8] McCloskey S, Jelinek J and Au K 2009 Method and system for determining shutter fluttering sequence US Patent 12/421,296
[9] Levin A, Sand P, Cho T, Durand F and Freeman W 2008 *ACM Transactions on Graphics (TOG)* vol 27 (ACM) p 71

Journal of Physics: Conference Series **386** (2012) 012002 doi:10.1088/1742-6596/386/1/012002

Non-heuristic reduction of the graph in graph-cut optimization

François Malgouyres[1] and Nicolas Lermé[2]

[1]Institut de Mathématiques de Toulouse, CNRS UMR 5219, Université de Toulouse, France
[2]LAGA, CNRS UMR 7539 and LIPN CNRS UMR 7030, Université de Paris 13, France

E-mail:
[1]Francois.Malgouyres@math.univ-toulouse.fr [2]nicolas.lerme@lipn.univ-paris13.fr

Abstract. During the last ten years, graph cuts had a growing impact in shape optimization. In particular, they are commonly used in applications of shape optimization such as image processing, computer vision and computer graphics. Their success is due to their ability to efficiently solve (apparently) difficult shape optimization problems which typically involve the perimeter of the shape. Nevertheless, solving problems with a large number of variables remains computationally expensive and requires a high memory usage since underlying graphs sometimes involve billion of nodes and even more edges. Several strategies have been proposed in the literature to improve graph-cuts in this regards. In this paper, we give a formal statement which expresses that a simple and local test performed on every node before its construction permits to avoid the construction of useless nodes for the graphs typically encountered in image processing and vision. A useless node is such that the value of the maximum flow in the graph does not change when removing the node from the graph. Such a test therefore permits to limit the construction of the graph to a band of useful nodes surrounding the final cut.

1. Introduction

Shape optimization is a field which has many practical applications. Among shape optimization methods, graph-cut methods have recently received a growing interest because of their ability to minimize pairwise Markov Random Fields of the form

$$E(u) = \beta \sum_{p\in\mathcal{P}} E_p(u_p) + \sum_{(p,q)\in(\mathcal{P}\times\mathcal{P})} E_{p,q}(u_p, u_q), \qquad \beta \in \mathbb{R}^+, \tag{1}$$

among $u \in \{0,1\}^{\mathcal{P}}$, for a set of pixels $\mathcal{P} \subset \mathbb{Z}^d$ and for a positive integer d. Indeed, when the terms $E_{p,q}(.)$ are submodular, there is a way to build a graph (see [1, 2]) such that if, for any $S \subset \mathcal{P}$, we denote by $u^S \in \{0,1\}^{\mathcal{P}}$ the binary partition of $\mathcal{P}$ defined for all $p \in \mathcal{P}$ by

$$u_p^S = \begin{cases} 1 & \text{, if } \; p \in S, \\ 0 & \text{, if } \; p \notin S, \end{cases} \tag{2}$$

we have

$$\mathrm{val}_{\mathcal{G}}(S) = E(u^S) + K, \tag{3}$$

for some (irrelevant in the current minimization context) constant $K \in \mathbb{R}$ and where $\mathrm{val}_{\mathcal{G}}(S)$ is the value of the cut defined by $S \subset \mathcal{P}$ (see next section for a precise definition).

Moreover, since (2) makes a one to one correspondence between cuts and elements of $\{0,1\}^{\mathcal{P}}$, we trivially obtain from (3) that to any minimum cut (min-cut) in $\mathcal{G}$ corresponds a minimizer of (1) and vice versa. Moreover, a min-cut can efficiently be computed with a maximum flow (max-flow) algorithm such as the one described in [3, 4]. The max flow is indeed the dual of the min-cut problem.

One drawback of graph-cuts is that the graph contains as many nodes as the number of pixels and that its number of edges behaves like

$$O\left(\sum_{p\in\mathcal{P}} \#\sigma(p)\right),$$

where $\#\sigma(p)$ is the number of neighbors of a pixel $p \in \mathcal{P}$ (see Section 2 for a detailed description of the graph). Such a memory requirement is a strong limitation when minimizing (1) for $\#\mathcal{P}$ large and/or for d large ($\#\sigma(p)$ typically grows like a radius raised to the power d...)

As a consequence, many strategies have been proposed to avoid the construction of the whole graph. We can find in the literature few methods that actually provide a global solution of the minimization problem [5, 6, 7]. However, most of he proposed methods are heuristics who may fail to accurately preserve thin structures [8, 9, 10, 11, 12]. The strategy developed in the current paper aims at building a thin band surrounding the final boundary between the object and the background. It shares this property with the other band-based methods (see [13, 10, 7, 9, 8]).

In this paper, we propose not to build any node satisfying a local test. This test is similar to the ones in [14, 13]. The nodes that fail to comply with the test are built in the reduced graph. The difference with the results stated in [14, 13] is that the test has been slightly modified in order to obtain a formal statement guaranteeing that the value of the max-flow in the reduced graph is equal to the value of the max-flow in the initial graph. As a consequence, any max-flow in the reduced graph is also a max-flow in the initial graph and the cut defined by this flow is a min-cut in the initial graph. In other words, despite the graph reduction, the max-flow algorithm still permits to build a minimizer of E.

The proof of the result presented in this paper as well as experiments are provided in [15] and will also be submitted for publication in a long version of the current paper.

2. Definition of the graph

The graph constructed in [1] has the form described below. This constitute the main motivation for considering this graph structure.

We consider two terminal nodes s and t and the set of nodes

$$\mathcal{V} = \mathcal{P} \cup \{s,t\}.$$

We consider a set of directed edges $\mathcal{E} \subset (\mathcal{V} \times \mathcal{V})$ such that $(\mathcal{V}, \mathcal{E})$ is a simple directed graph. We also assume that for every $p \in \mathcal{P}$,

$$(p,s) \notin \mathcal{E} \text{ and } (t,p) \notin \mathcal{E}.$$

We denote the neighbors of any nodes $p \in \mathcal{V}$ by

$$\sigma_{\mathcal{E}}(p) = \{q \in \mathcal{V}, (p,q) \in \mathcal{E} \text{ or } (q,p) \in \mathcal{E}\}.$$

We define the capacities as a mapping $c : (\mathcal{V} \times \mathcal{V}) \to \mathbb{R}^+$ and denote the capacity of any edge $(p,q) \in (\mathcal{V} \times \mathcal{V})$ by

$$c_{p,q} \geq 0.$$

Notice that usually, the capacities are only defined over $\mathcal{E}$. We extend their definition to $(\mathcal{V}\times\mathcal{V})$ in order to simplify notations. This extension is performed by setting

$$c_{p,q} = 0, \text{ whenever } (p,q) \notin \mathcal{E}. \tag{4}$$

We assume, without loss of generality (see [2, 1]), that capacities are such that for every $p \in \mathcal{P}$

$$c_{s,p} \neq 0 \quad \Rightarrow \quad c_{p,t} = 0.$$

We therefore sum up the capacities of the edges linked to the terminal nodes and set for all $p \in \mathcal{P}$

$$c_p = c_{s,p} - c_{p,t}.$$

For any $S \subset \mathcal{P}$, we denote by $\text{val}_{\mathcal{G}}(S)$ the value of the s-t cut

$$(S \cup \{s\}, (\mathcal{P} \setminus S) \cup \{t\}),$$

in $\mathcal{G}$.

We also define flows in the graph as any mapping $f : (\mathcal{V}\times\mathcal{V}) \to \mathbb{R}^+$ satisfying both the capacity constraints

$$0 \leq f_{p,q} \leq c_{p,q} \quad , \text{ for all } (p,q) \in (\mathcal{V}\times\mathcal{V}), \tag{5}$$

and the flow conservation

$$\sum_{q\in\sigma_{\mathcal{E}}(p)} f_{q,p} = \sum_{q\in\sigma_{\mathcal{E}}(p)} f_{p,q}, \qquad \text{for all } p \in \mathcal{P}. \tag{6}$$

As usual, the value of the flow f in $\mathcal{G}$ is defined by

$$\text{val}_{\mathcal{G}}(f) = \sum_{p\in\sigma_{\mathcal{E}}(s)} f_{s,p}. \tag{7}$$

Notice that we use the same notation for the value of a flow and the value of a s-t cut in $\mathcal{G}$. This abuse of notation will never be ambiguous, once in context.

We call max-flow any solution f^* of the linear program

$$\begin{cases} \max_f \quad \text{val}_{\mathcal{G}}(f), \\ \text{subject to (5) and (6).} \end{cases}$$

In the sequel, we also consider a subset $B \subset \mathbb{Z}^d$ and assume that B and $\mathcal{G}$ are such that

$$\forall p \in \mathcal{P}, \quad (\sigma_{\mathcal{E}}(p) \cap \mathcal{P}) \; \subset \; B_p, \tag{8}$$

where

$$B_p = \{p+q, q \in B\}. \tag{9}$$

In practice, we typically think of B as a ball centered at the origin. In such a case, (8) means that neighbors in the graph $\mathcal{G}$ are close to each other in $\mathbb{Z}^d$.

3. Main result of the paper

Theorem 1 *Let $\mathcal{G}$ be a graph as defined in Section 2, let B satisfy (8) and let us assume that $p \in \mathcal{P}$ satisfies*

$$\begin{cases} \textit{either} & \forall q \in B_p, \quad c_q \geq \sum_{q' \notin B_p} c_{q,q'}, \\ \textit{or} & \forall q \in B_p, \quad c_q \leq -\sum_{q' \notin B_p} c_{q',q}. \end{cases} \tag{10}$$

Then, there exists a max-flow f in $\mathcal{G}$ such that

$$\forall q \in \sigma_{\mathcal{E}}(p), \qquad f_{p,q} = f_{q,p} = 0.$$

As a consequence, removing the node p from the graph $\mathcal{G}$ does not modify its max-flow value.

Algorithmically, the above theorem guarantees that we can test every node, during the graph construction, before it is added to the graph. If the node satisfies (10), it is not useful to the max-flow evaluation. It can therefore be removed from the graph $\mathcal{G}$ without modifying its max-flow value.

Let us describe the meaning of the test corresponding to the upper line of (10). For every node q in B_p

- either q is in the interior of B_p in which case, for all $q' \notin B_p$, $c_{q,q'} = 0$ and the test simply becomes $c_q \geq 0$.
- or q is on the border of B_p and then the test takes the form $c_q \geq \sum_{q' \notin B_p} c_{q,q'} \geq 0$ and is generally more difficult to satisfy.

Notice also that a similar theorem with the test

$$\begin{cases} \text{either} & \forall q \in B_p, \quad c_q \geq \sum_{q' \in \sigma_{\mathcal{E}}(q)} c_{q,q'}, \\ \text{or} & \forall q \in B_p, \quad c_q \leq -\sum_{q' \in \sigma_{\mathcal{E}}(q)} c_{q',q}, \end{cases}$$

is easy to obtain. The two results should not be confused.

Let us now describe how to build the minimizer of (1), once the minimum cut

$$\Big(S \cup \{s\}\ ,\ ((\mathcal{P} \setminus \{p\}) \setminus S) \cup \{t\}\Big),$$

where $S \subset (\mathcal{P} \setminus \{p\})$, has been computed in the graph from which the node p satisfying (10) has been removed. The minimizer $u^* \in \{0,1\}^{\mathcal{P}}$ of (1) is defined by

$$u_r^* = \begin{cases} 1 & \text{, if } r = p \text{ and } \forall q \in B_p, \quad c_q \geq \sum_{q' \notin B_p} c_{q,q'}, \\ 0 & \text{, if } r = p \text{ and } \forall q \in B_p, \quad c_q \leq -\sum_{q' \notin B_p} c_{q',q}, \\ 1 & \text{, if } r \neq p \text{ and } r \in S, \\ 0 & \text{, if } r \neq p \text{ and } r \notin S. \end{cases}$$

The proof of the above theorem as well as some experiments illustrating the situations in which it can successfully be applied are provided in [16].

References

[1] Kolmogorov V and Zabih R 2004 *IEEE Transactions on Pattern Analysis and Machine Intelligence* **26** 147–159

[2] Picard J and Ratliff H 1975 *Networks* **5** 357–370

[3] Boykov Y and Kolmogorov V 2004 *IEEE Transactions on Pattern Analysis and Machine Intelligence* **26** 1124–1137

[4] Hochbaum D S 2008 *Operation research* **56** 992–1009

[5] Delong A and Boykov Y 2008 *CVPR* pp 1–8

Journal of Physics: Conference Series **386** (2012) 012002 doi:10.1088/1742-6596/386/1/012002

[6] Strandmark P and Kahl F 2010 *CVPR* pp 2085–2092
[7] Lempitsky V and Boykov Y 2007 *CVPR* pp 1–8
[8] Lombaert H, Sun Y, Grady L and Xu C 2005 *ICCV* vol 1 pp 259–265
[9] Sinop A and Grady L 2006 *MICCAI* vol 2 pp 896–903
[10] Kohli P, Lempitsky V and Rother C 2010 *DAGM* pp 242–251
[11] Li Y, Sun J, Tang C and Shum H 2004 *ACM Transactions on Graphics* **23** 303–308
[12] Stawiaski J, Decencière E and Bidault F 2007 *ISMM* pp 349–360
[13] Lermé N and Malgouyres F 2011 A reduction method for graph cut optimization Tech. Rep. hal-00606921 CCSD
[14] Lermé N, Malgouyres F and Létocart L 2010 *ICIP* pp 3045–3048
[15] Lermé N 2011 *Réduction de graphes et application la segmentation de tumeurs pulmonaires* Ph.D. thesis Université Paris 13
[16] Malgouyres F and lermé N 2012 A non-heuristic reduction method for graph cut optimization Tech. Rep. hal-00692464 CCSD preprint

2nd Int. Workshop on New Computational Methods for Inverse Problems (NCMIP 2012) IOP Publishing
Journal of Physics: Conference Series **386** (2012) 012003 doi:10.1088/1742-6596/386/1/012003

Proximal Splitting Derivatives for Risk Estimation

C. Deledalle[1], S. Vaiter[1], G. Peyré[1], J. Fadili[2], C. Dossal[3]
[1] CEREMADE, CNRS-Université Paris-Dauphine
[2] GREYC, CNRS-ENSICAEN-Université de Caen
[3] IMB, CNRS-Université Bordeaux 1

Abstract. This paper develops a novel framework to compute a projected Generalized Stein Unbiased Risk Estimator (GSURE) for a wide class of sparsely regularized solutions of inverse problems. This class includes arbitrary convex data fidelities with both analysis and synthesis mixed $\ell^1 - \ell^2$ norms. The GSURE necessitates to compute the (weak) derivative of a solution w.r.t. the observations. However, as the solution is not available in analytical form but rather through iterative schemes such as proximal splitting, we propose to iteratively compute the GSURE by differentiating the sequence of iterates. This provides us with a sequence of differential mappings, which, hopefully, converge to the desired derivative and allows to compute the GSURE. We illustrate this approach on total variation regularization with Gaussian noise to automatically select the regularization parameter.

1. Introduction

This paper focuses on unbiased estimation of the ℓ^2-risk of recovering an image $f_0 \in \mathbb{R}^N$ from low-dimensional noisy observations $y = \Phi f_0 + w$, where $w \sim \mathcal{N}(0, \sigma^2 \mathrm{Id}_P)$. The linear bounded imaging operator $\Phi : \mathbb{R}^N \to \mathcal{Y} = \mathbb{R}^P$ entails loss of information so that $P < N$ or is rank-deficient for $P = N$, and the recovery problem is typically ill-posed.

In the following we denote $f(y) \in \mathbb{R}^N$ the estimator of f_0 from the observations $y \in \mathcal{Y}$. More specifically, we consider an estimator $f(y)$ defined as a function of coefficients $x(y) \in \mathcal{X}$ (where $\mathcal{X}$ is a suitable finite-dimensional Hilbert space) that solves $x(y) \in \operatorname{argmin}_{x \in \mathcal{X}} E(x, y)$ where the set of minimizers is nonempty. Here $E(x, y)$ is an energy functional parameterized by the observations $y \in \mathcal{Y}$. In some cases (e.g. total variation regularization), one directly has $f(y) = x(y)$, but for sparse regularization in a redundant synthesis dictionary, the latter maps coefficients $x(y)$ to images $f(y)$. We then make a distinction between $f(y)$ and $x(y)$ in the following.

This work proposes a versatile approach for unbiased risk estimation in the case where $x(y)$ is computed by proximal splitting algorithms. These methods have become extremely popular to solve inverse problems with convex non-smooth regularizations, e.g. those encountered in the sparsity field.

2. Previous Works

Unbiased Risk Estimation. The SURE [1] is an unbiased ℓ^2-risk estimator. For denoising $\Phi = \mathrm{Id}$, it provides an unbiased estimate SURE(y) of the risk $\mathbb{E}(\|f(y) - f_0\|^2)$ that depends solely on y, without prior knowledge of f_0. This can prove very useful for objective choice of parameters that minimize the recovery risk of f_0. A generalized SURE (GSURE) has been developed for noise models within the multivariate canonical exponential family [2]. In the context of inverse problems, this allows to compute the projected risk $\mathbb{E}(\|\Pi(f(y) - f_0)\|^2)$ where

Π is the orthogonal projector on $\ker(\Phi)^\perp$. Similar GSURE versions have been proposed for Gaussian noise and special regularizations or/and inverse problems, e.g. [3, 4].

Unbiased Estimation of the Degrees of Freedom. A prerequisite to compute the SURE or GSURE is an unbiased estimate of the degrees of freedom $\mathrm{df}(y)$. Roughly speaking, for overdetermined linear models, $\mathrm{df}(y)$ is the number of free parameters in modeling $f(y)$ from y. There are situations where $\mathrm{df}(y)$ can be estimated in closed-form from $f(y)$. This occurs e.g. in synthesis ℓ^1 regularization, as established in the overdetermined case in [5], and extended to the general setting in [6]. When no closed-form is available, $\mathrm{df}(y)$ can be estimated using data perturbation and Monte-Carlo integration, see e.g. [7]. Alternatively, an estimate can be obtained by formally differentiating the sequence of iterates provided by an algorithm that converges to $f(y)$. As proposed initially by [4] and refined in [8], this allows to compute the GSURE of sparse synthesis regularization.

3. Differentiating Proximal Splitting Schemes

Due to obvious space limitations, we only describe here the derivative of a primal-dual proximal splitting algorithm. But the same idea remains valid for other splitting schemes as it is described in a longer version of this paper [9].

3.1. Proximal Operator

The proximal operator associated to a proper lower semi-continuous (lsc) and convex function $x \mapsto G(x, y)$ is

$$\mathrm{Prox}_G(x, y) = \underset{z}{\mathrm{argmin}} \ \frac{1}{2}\|x - z\|^2 + G(z, y).$$

A function for which $\mathrm{Prox}_G(x, y)$ can be computed in closed-form is dubbed simple. A distinctive property of $\mathrm{Prox}_G(\cdot, y)$ that plays a central role in the sequel is that its is a 1-Lipschitz mapping. When y is fixed, we will denote $\mathrm{Prox}_G(x)$ instead of $\mathrm{Prox}_G(x, y)$ to lighten the notation.

The Legendre-Fenchel conjugate of G is $G^*(z, y) = \max_x \langle x,\, z\rangle - G(x, y)$. A useful proximal calculus rule is Moreau's identity: $x = \mathrm{Prox}_{\tau G^*}(x, y) + \tau\, \mathrm{Prox}_{G/\tau}(x/\tau, y),\ \tau > 0$.

3.2. Primal-Dual Splitting

Proximal splitting schemes can be used to solve the large class of variational problems

$$x(y) \in \underset{x \in \mathcal{X}}{\mathrm{argmin}} \ E(x, y) = H(x, y) + G(Kx, y) \ , \tag{1}$$

where both $x \mapsto H(x, y)$ and $u \mapsto G(u, y)$ are proper, lsc, convex and simple functions, and $K : \mathcal{X} \to \mathcal{U}$ is a bounded linear operator.

The primal-dual relaxed Arrow-Hurwicz algorithm as proposed in [10] to solve (1) reads

$$\begin{aligned} u^{(\ell+1)} &= \mathrm{Prox}_{\sigma H^*}(U^{(\ell)}) \quad \text{where} \quad U^{(\ell)} = u^{(\ell)} + \sigma K \tilde{x}^{(\ell)}, \\ x^{(\ell+1)} &= \mathrm{Prox}_{\tau G}(X^{(\ell)}) \quad \text{where} \quad X^{(\ell)} = x^{(\ell)} - \tau K^* u^{(\ell)}, \\ \tilde{x}^{(\ell+1)} &= x^{(\ell+1)} + \theta(x^{(\ell+1)} - x^{(\ell)}) \end{aligned} \tag{2}$$

where $u^{(\ell)} \in \mathcal{U}$, $x^{(\ell)} \in \mathcal{X}$ and $\tilde{x}^{(\ell)} \in \mathcal{X}$. The parameters $\sigma > 0, \gamma > 0$ are chosen such that $\sigma\gamma\|K\|^2 < 1$, and $\theta \in [0, 1]$ to ensure convergence of $x^{(\ell)}$ toward a global minimizer of (1). $\theta = 0$ corresponds to the Arrow-Hurwicz algorithm, and for $\theta = 1$ a convergence rate of $O(1/\ell)$ was established on the restricted duality gap [10].

For any vector $\delta \in \mathcal{Y}$, our goal is to compute the derivatives $\xi^{(\ell)} = \partial x^{(\ell)}(y)[\delta]$, $\upsilon^{(\ell)} = \partial u^{(\ell)}(y)[\delta]$ and $\tilde{\xi}^{(\ell)} = \partial \tilde{x}^{(\ell)}(y)[\delta]$. Using the chain rule, the sequence of derivatives then reads

$$\begin{aligned} \upsilon^{(\ell+1)} &= \mathcal{H}_1^{(\ell)}(\Upsilon^{(\ell)}) + \mathcal{H}_2^{(\ell)}(\delta) \quad \text{where} \quad \Upsilon^{(\ell)} = \upsilon^{(\ell)} + \sigma K \tilde{\xi}^{(\ell)}, \\ \xi^{(\ell+1)} &= \mathcal{G}_1^{(\ell)}(\Xi^{(\ell)}) + \mathcal{G}_2^{(\ell)}(\delta) \quad \text{where} \quad \Xi^{(\ell)} = \xi^{(\ell)} - \tau K^* \upsilon^{(\ell)}, \\ \tilde{\xi}^{(\ell+1)} &= \xi^{(\ell+1)} + \theta(\xi^{(\ell+1)} - \xi^{(\ell)}) \end{aligned} \tag{3}$$

where we have defined the following linear mappings for $k = 1, 2$ with ∂_k the derivative w.r.t. the k-th argument

$$\mathcal{H}_k^{(\ell)}(\cdot) = \partial_k \operatorname{Prox}_{\sigma H^*}(U^{(\ell)}, y)[\cdot] \quad \text{and} \quad \mathcal{G}_k^{(\ell)}(\cdot) = \partial_k \operatorname{Prox}_{\tau G}(X^{(\ell)}, y)[\cdot].$$

3.3. Discussion on convergence issues

One has to be aware that given that the proximal mappings are not necessarily differentiable everywhere, its differential is actually set-valued. Therefore, one should appeal to involved tools from non-smooth analysis to make the above statements rigorous. We prefer not to delve into these technicalities for the lack of space.

Another major issue is to theoretically ensure the existence of a proper sequence $\xi^{(\ell)}$ that converges toward $\partial x(y)[\delta]$. Regarding existence, $\operatorname{Prox}_G(\cdot, y)$ is a 1-Lipschitz mapping of its first argument. Furthermore, in all the considered application, $\operatorname{Prox}_G(x, \cdot)$ is also Lipschitz with respect to its second argument. If one starts at an appropriate initialization, by induction, $y \mapsto x^{(\ell)}(y)$ is also Lipschitz, hence differentiable almost everywhere. As far as convergence is concerned, this remains an open question in the general case, and we believe this would necessitate intricate arguments from non-smooth and variational analysis. This is left to future research.

4. Risk Estimator

Projected GSURE. Recall that $\Pi = \Phi^*(\Phi\Phi^*)^+\Phi$ is the orthogonal projector on $\ker(\Phi)^\perp = \operatorname{Im}(\Phi^*)$, where $^+$ stands for the Moore-Penrose pseudo-inverse. Let $\mu(y) = \Pi f(y)$ the projected estimator of Πf_0. While $f(y)$ is not necessarily uniquely defined, we assume that $\mu(y)$ is unambiguously defined as a single-valued mapping of the observation y. This can be ensured under a strict convexity condition on H or G in (1) (see e.g. example (5)).

Let $\mu_0(y) = \Phi^*(\Phi\Phi^*)^+ y$ the maximum likelihood estimator. By generalizing the projected GSURE of [3] to any linear operator Φ, we have

$$\operatorname{GSURE}(y) = \|\mu_0(y) - \mu(y)\|^2 - \sigma^2 \operatorname{tr}((\Phi\Phi^*)^+) + 2\sigma^2 \operatorname{div}((\Phi\Phi^*)^+ \Phi f(y))$$

where $\operatorname{div}(g)(y) = \operatorname{tr}(\partial g(y))$ is the divergence of the mapping $g : \mathcal{Y} \to \mathcal{Y}$. Under weak differentiability of $y \mapsto \mu(y)$, one can prove that the GSURE is an unbiased estimate of the risk on $\operatorname{Im}(\Phi^*)$, i.e. $\mathbb{E}_w(\operatorname{GSURE}(y)) = \mathbb{E}_w(\|\Pi f_0 - \mu(y)\|^2)$.

Iterative Numerical Computation. One of the bottlenecks in calculating the GSURE(y) is to efficiently compute the divergence term. Using the Jacobian trace formula of the divergence, it can be easily seen that

$$\operatorname{div}((\Phi\Phi^*)^+ \Phi f(y)) = \mathbb{E}_z(\langle \partial f(y)[z], \mu_0(z) \rangle) \approx \frac{1}{k} \sum_{i=1}^{k} \langle \partial f(y)[z_i], \mu_0(z_i) \rangle \tag{4}$$

where $z \sim \mathcal{N}(0, \operatorname{Id}_P)$ and z_i are k realizations of z. Since $f(y)$ and its iterates $f^{(\ell)}(y)$ are defined as explicit functions of $x(y)$ and $x^{(\ell)}(y)$ (see Section 5 for a detailed example), the GSURE(y) can in turn be iteratively estimated by plugging $\partial x^{(\ell)}(y)[z_i]$ provided by (3) into (4).

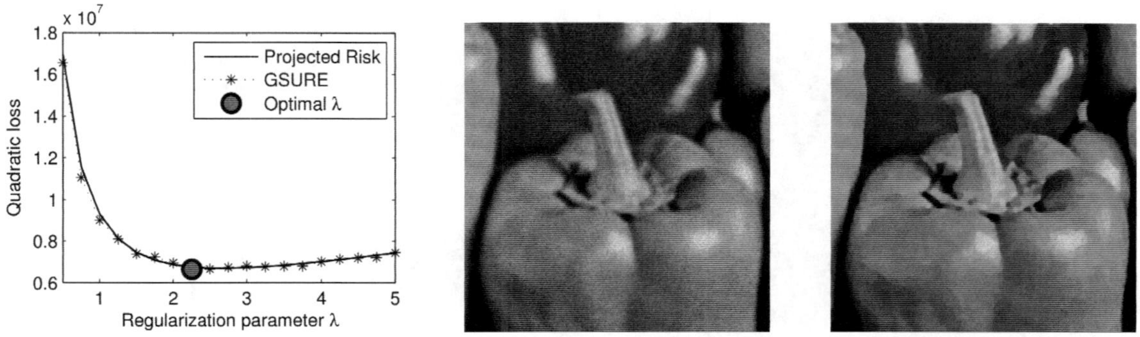

(a) (b) y (c) $f(y)$ at the optimal λ

Figure 1. (a) Projected risk and its GSURE estimate[1] (b) y. (c) $f(y)$ at the optimal λ.

5. Numerical Results

Total variation regularization of linear inverse problems amounts to solving

$$f(y) \in \underset{f}{\operatorname{argmin}} \ \frac{1}{2}\|\Phi f - y\|^2 + \lambda \|\nabla f\|_1 \ , \tag{5}$$

where $\nabla f \in \mathbb{R}^{N \times 2}$ is a discrete gradient. The $\ell^1 - \ell^2$ norm of a vector field $t = (t_i)_{i=1}^N \in \mathbb{R}^{N \times 2}$, with $t_i \in \mathbb{R}^2$, is defined as $\|t\|_1 = \sum_i \|t_i\|$. Problem (5) is a special instance of (1) letting $x = f$, $H(x, y) = 0, \forall (x, y)$ and

$$K(x) = (\Phi x, \nabla x) \quad \text{and} \quad \forall\, u = (s, t) \in \mathbb{R}^P \times \mathbb{R}^{N \times 2}, \quad G(u, y) = \frac{1}{2}\|s - y\|^2 + \lambda \|t\|_1 \ .$$

Separability of G in s and t entails that

$$\operatorname{Prox}_{\tau G}(u, y) = ((1 - \tau)s + \tau y, T_{\lambda\tau}(t)) \ ,$$

where $T_\rho, \rho > 0$, is the component-wise $\ell^1 - \ell^2$ soft-thresholding, defined for $i = 1, \ldots, N$ as

$$T_\rho(t)_i = \max(0, 1 - \rho/\|t_i\|)t_i \quad \text{and} \quad \partial T_\rho(t)[\delta_t]_i = \begin{cases} 0 & \text{if} \ \ \|t_i\| \leqslant \rho \\ \delta_{t,i} - \frac{\rho}{\|t_i\|} P_{t_i}(\delta_{t,i}) & \text{otherwise} \end{cases} ,$$

where P_α is the orthogonal projector on $\alpha^\perp$ for $\alpha \in \mathbb{R}^2$, and $T_\rho(t)_i$ although not differentiable on the sphere $\{t_i : \|t_i\| = \rho\}$, is directionally differentiable there. Therefore

$$\partial_1 \operatorname{Prox}_{\tau G}(u, y)[\delta_s, \delta_t] = ((1 - \tau)\delta_s, \partial T_{\lambda\tau}(\delta_t)) \quad \text{and} \quad \partial_2 \operatorname{Prox}_{\tau G}(u, y)[\delta_y] = (\tau \delta_y, 0) \ .$$

Fig. 1 depicts an application of our GSURE to adjust the value of λ optimizing the recovery for a deblurring problem, where $\Phi \in \mathbb{R}^{N \times N}$ is a convolution matrix (Gaussian kernel of width 2 px), $\sigma = 10$ (for an image f_0 with a range $[0, 255]$). For each value of λ in the tested range, GSURE(y) is computed for a single realization of y using (4) with $k = 4$ realizations z_i.

References

[1] Stein C 1981 *The Annals of Statistics* **9** 1135–1151
[2] Eldar Y C 2009 *IEEE Transactions on Signal Processing* **57** 471–481
[3] Pesquet J C, Benazza-Benyahia A and Chaux C 2009 *IEEE Transactions on Signal Processing* **57** 4616–4632
[4] Vonesch C, Ramani S and Unser M 2008 *ICIP* (IEEE) pp 665–668
[5] Zou H, Hastie T and Tibshirani R 2007 *The Annals of Statistics* **35** 2173–2192
[6] Kachour M, Dossal C, Fadili J, Peyré G and Chesneau C 2011 The degrees of freedom of penalized l1 minimization Tech. rep. Preprint Hal-00638417
[7] Ye J 1998 *Journal of the American Statistical Association* **93** 120–131
[8] Giryes R, Elad M and Eldar Y 2011 *Applied and Computational Harmonic Analysis* **30** 407–422
[9] Deledalle C, Vaiter S, Peyré G, Fadili J and Dossal C 2012 Proximal splitting derivatives for unbiased risk estimation Tech. rep. Preprint Hal-00662719 URL `http://hal.archives-ouvertes.fr/hal-00670213`
[10] Chambolle A and Pock T 2011 *Journal of Mathematical Imaging and Vision* **40** 120–145

[1] Without impacting the optimal choice of λ, the two curves have been vertically shifted for visualization purposes.

Data completion problems solved as Nash games

A Habbal[1] and M Kallel[2]

[1] Laboratoire J.A.Dieudonné, Université de Nice & INRIA, France
[2] Laboratoire LAMSIN-ENIT & IPEIT, Université de Tunis, Tunisia

E-mail: habbal@polytech.unice.fr and moez.kallel@ipeit.rnu.tn

Abstract.
The Cauchy problem for an elliptic operator is formulated as a two-player Nash game.
Player (1) is given the known Dirichlet data, and *uses as strategy variable the Neumann condition* prescribed over the inaccessible part of the boundary.
Player (2) is given the known Neumann data, and *plays with the Dirichlet condition* prescribed over the inaccessible boundary.
The two players solve in parallel the associated Boundary Value Problems. Their respective objectives involve the *gap between the non used Neumann/Dirichlet known data and the traces of the BVP's solutions* over the accessible boundary, and are *coupled through a difference term.* We prove the existence of a unique Nash equilibrium, which turns out to be the reconstructed data when the Cauchy problem has a solution. We also prove that the completion algorithm is stable with respect to noise, and present two 3D experiments which illustrate the efficiency and stability of our algorithm.

1. Introduction

Let be Ω a bounded open domain in $\mathbb{R}^d$ ($d = 2, 3$) with a sufficiently smooth boundary $\partial\Omega$ composed of two connected disjoint components Γ_c and Γ_i. We consider the following elliptic Cauchy problem :

$$\left\{\begin{array}{rcll} \nabla.(k\nabla u) & = & 0 & \text{in } \Omega \\ u & = & f & \text{on } \Gamma_c \\ k\nabla u.\nu & = & \Phi & \text{on } \Gamma_c \end{array}\right. \tag{1}$$

The parameters k, f and Φ are given functions, ν is the unit outward normal vector on the boundary. The Dirichlet data f and the Neumann data Φ are the so-called Cauchy data, which are known on the accessible part Γ_c of the boundary $\partial\Omega$ and the unknown field u is the Cauchy solution. The Cauchy problem is also known as a data completion problem, where the data to be recovered, or missing data, are $u_{|\Gamma_i}$ and $k\nabla u.\nu_{|\Gamma_i}$, which are determined as soon as one knows u in the whole Ω.

Completion/Cauchy problems are known to be severely ill-posed (Hadamard's), and computationally challenging.

Many authors ([1]) have formulated the Cauchy problem (1) as an optimal control one, as follows :

Let us assume that $(\Phi, f) \in H^{-\frac{1}{2}}(\Gamma_c) \times H^{\frac{1}{2}}(\Gamma_c)$. For given $\eta \in H^{-\frac{1}{2}}(\Gamma_i)$ and $\tau \in H^{\frac{1}{2}}(\Gamma_i)$, let us define $u_1(\eta)$ and $u_2(\tau)$ as the unique solutions in $H^1(\Omega)$ of the following elliptic boundary value problems :

$$(SP1)\left\{\begin{array}{rcll} \nabla.(k\nabla u_1) & = & 0 & \text{in } \Omega \\ u_1 & = & f & \text{on } \Gamma_c \\ k\nabla u_1.\nu & = & \eta & \text{on } \Gamma_i \end{array}\right. \qquad (SP2)\left\{\begin{array}{rcll} \nabla.(k\nabla u_2) & = & 0 & \text{in } \Omega \\ u_2 & = & \tau & \text{on } \Gamma_i \\ k\nabla u_2.\nu & = & \Phi & \text{on } \Gamma_c \end{array}\right. \tag{2}$$

The optimization problem amounts to minimize, among all pairs $(\eta, \tau) \in H^{-\frac{1}{2}}(\Gamma_i) \times H^{\frac{1}{2}}(\Gamma_i)$, the following "Neumann-gap" cost :

$$\mathcal{J}_1(\eta,\tau) = \mathrm{J}_1(\eta,\tau,u_1(\eta),u_2(\tau)) \;=\; \frac{1}{2}\|k\nabla u_1.\nu - \Phi\|^2_{H^{-\frac{1}{2}}(\Gamma_c)} + \frac{1}{2}\|u_1 - u_2\|^2_{L^2(\Gamma_i)}. \tag{3}$$

It is known that *when the Cauchy problem (1) has a solution*, then solving it is equivalent to solving the minimization problem

$$\min_{(\eta,\tau)\in H^{-\frac{1}{2}}(\Gamma_i)\times H^{\frac{1}{2}}(\Gamma_i)} \mathcal{J}_1(\eta,\tau). \tag{4}$$

The same conclusions above hold when a "Dirichlet-gap" cost is considered :

$$\mathcal{J}_2(\eta,\tau) = \mathrm{J}_2(\eta,\tau,u_1(\eta),u_2(\tau)) \;=\; \frac{1}{2}\|u_2 - f\|^2_{L^2(\Gamma_c)} + \frac{1}{2}\|u_1 - u_2\|^2_{L^2(\Gamma_i)}. \tag{5}$$

Formulated in the game theory vocabulary, the Neumann and Dirichlet controls η and τ do cooperate to minimize either the Neumann-gap or the Dirichlet-gap costs. These two controls could as well cooperatively minimize any convex combination of the two costs $\mathcal{J}_1$ and $\mathcal{J}_2$.

Now, the fields $u_1(\eta)$ and $u_2(\tau)$ are aiming at the fulfillment of a possibly antagonistic goals, namely minimizing the Neumann gap $\|k\nabla u_1.\nu - \Phi\|_{H^{-\frac{1}{2}}(\Gamma_c)}$ and the Dirichlet gap $\|u_2 - f\|_{L^2(\Gamma_c)}$. This antagonism is intimately related to Hadamard's ill-posedness character of the Cauchy problem, and rises as soon as one requires that u_1 and u_2 coincide, which is exactly what the coupling term $\|u_1 - u_2\|_{L^2(\Gamma_i)}$ is for. Thus, one may think of an iterative process which minimizes in a smart fashion the three terms, namely Neumann-Dirichlet-Coupling terms.

Let us define the following two costs : for any $\eta \in H^{-\frac{1}{2}}(\Gamma_i)$ and $\tau \in H^{\frac{1}{2}}(\Gamma_i)$,

$$J_1(\eta,\tau) = \frac{1}{2}\|k\nabla u_1.\nu - \Phi\|^2_{H^{-\frac{1}{2}}(\Gamma_c)} + \frac{\alpha}{2}\|u_1 - u_2\|^2_{H^{\frac{1}{2}}(\Gamma_i)}, \tag{6}$$

$$J_2(\eta,\tau) = \frac{1}{2}\|u_2 - f\|^2_{L^2(\Gamma_c)} + \frac{\alpha}{2}\|u_1 - u_2\|^2_{H^{\frac{1}{2}}(\Gamma_i)}. \tag{7}$$

where the fields $u_1(\eta)$ and $u_2(\tau)$ are the unique solutions to (SP1) and (SP2), respectively and α is a given positive parameter (e.g. $\alpha = 1$).

We shall say that there are two players, referred to as player 1 or Neumann-gap, and player 2 or Dirichlet-gap. Player 1 controls the strategy variable η, and player 2 controls the strategy variable τ. Each of the two players tries to minimize its own cost, namely J_1 for player 1, and J_2 for player 2. As classical, the fact that each player controls only his own strategy, while there is a strong dependance of each player's cost on the joint strategies (η, τ) justifies the use of the game theory framework (and terminology), a natural setting which may be used to formulate the negotiation between these two costs.

In order to be consistent with the initial formulation of the Cauchy problem, the relevant game theoretic framework to deal with is a static with complete information one. In this case, a commonly used solution concept (roughly speaking, in the game vocabulary, a rational and stable one) is the one of Nash Equilibria. Our main results are the following :

Proposition 1 *Consider the Nash game defined by the two costs (6) and (7).*

(i) The partial mapping $\eta \to J_1(\eta,\tau)$ *(respectively* $\tau \to J_2(\eta,\tau)$*) is a quadratic strictly convex and coercive functional over* $H^{-\frac{1}{2}}(\Gamma_i)$ *(resp.* $H^{\frac{1}{2}}(\Gamma_i)$*).*

(ii) There always exists a unique Nash equilibrium $(\eta_N, \tau_N) \in H^{-\frac{1}{2}}(\Gamma_i) \times H^{\frac{1}{2}}(\Gamma_i)$ *. It is also the minimum of* $L(\eta,\tau) = \frac{1}{2}\|k\nabla u_1.\nu - \Phi\|^2_{H^{-\frac{1}{2}}(\Gamma_c)} + \frac{1}{2}\|u_2 - f\|^2_{L^2(\Gamma_c)} + \frac{\alpha}{2}\|u_1 - u_2\|^2_{H^{\frac{1}{2}}(\Gamma_i)}$.

(iii) When the Cauchy problem has a solution u, then $u_1(\eta_N) = u_2(\tau_N) = u$, and (η_N, τ_N) are the missing data, namely $\eta_N = k\nabla u.\nu_{|\Gamma_i}$ and $\tau_N = u_{|\Gamma_i}$.

Proposition 2 *Assume there exists a unique Cauchy solution $u \in H^1(\Omega)$ for a given compatible pair of data $(f, \Phi) \in H^{\frac{1}{2}}(\Gamma_c) \times H^{-\frac{1}{2}}(\Gamma_c)$. Let $(f^\delta, \Phi^\delta) \in H^{\frac{1}{2}}(\Gamma_c) \times H^{-\frac{1}{2}}(\Gamma_c)$ be any sequence of noisy data such that*

$$\|f^\delta - f\|^2_{H^{\frac{1}{2}}(\Gamma_c)} + \|\Phi^\delta - \Phi\|^2_{H^{-\frac{1}{2}}(\Gamma_c)} \leq \delta^2.$$

Then, the Nash game with the costs J_1^δ and J_2^δ, defined for the noisy data, has a unique Nash equilibrium $(\eta_N^\delta, \tau_N^\delta) \in H^{-\frac{1}{2}}(\Gamma_i) \times H^{\frac{1}{2}}(\Gamma_i)$ which strongly converges, as $\delta \to 0$, to the Cauchy missing data $(k\nabla u.\nu_{|\Gamma_i}, u_{|\Gamma_i})$.

Moreover, the fields $u_1^\delta(\eta_N^\delta)$ and $u_2^\delta(\tau_N^\delta)$, which solve respectively (SP1) and (SP2) for the noisy data, strongly converge in $H^1(\Omega)$ to the Cauchy solution u.

The proofs of the propositions lie on the classical Nash existence theorem, and on the not less classical Sobolev traces theorems and a priori estimates for elliptic BVP. The above results hold as well when considering as coupling term the difference $\|k\nabla u_1.\nu - k\nabla u_2.\nu\|_{H^{-\frac{1}{2}}(\Gamma_i)}$.

From the computational viewpoint, in [3] the authors propose an alternating minimization algorithm to compute the Nash Equilibrium by means of the following iterative process : given an initial (η^0, τ^0) ;

$$\begin{cases} \eta^{(k+1)} = \text{argmin}_\eta \{J_1(\eta, \tau^{(k)}) + \frac{\beta}{2} \int_{\Gamma_i} (\eta - \eta^{(k)})^2 d\Gamma_i\}, \\ \tau^{(k+1)} = \text{argmin}_\tau \{J_2(\eta^{(k)}, \tau) + \frac{\beta}{2} \int_{\Gamma_i} (\tau - \tau^{(k)})^2 d\Gamma_i\}, \end{cases} \tag{8}$$

where β is a given positive parameter (e.g. $\beta = 1$).

In the cited reference, the convergence of the alternating algorithm above is proved, under suitable assumptions which also hold in our case, see Proposition 1.

2. Two Numerical Experiments

The involved boundary value problems are solved by means of the classical dual Raviart-Thomas mixed finite element method, in order to get accurate approximations of the normal derivatives (which intervene as well in the adjoint state equations we used to compute the costs gradients).

2.1. Non-radial 3D test-case

We consider a thick spherical shell domain Ω with boundary components Γ_i and Γ_c, which are two spheres both centered at $(0, 0, 0)$ and with radii given by respectively $R_i = 0.6$ and $R_c = 1$.

The exact solution to the Cauchy problem is given by

$$u(x, y, z) = \frac{1}{\sqrt{(x + 0.2)^2 + y^2 + z^2}}. \tag{9}$$

provided that the Cauchy data f and Φ are defined as respectively the trace and normal derivative of the u above over the sphere Γ_c. Remark that the function u given by (9) is the non-radial solution of $\Delta u = \delta_{X_0}$, where the source term is now $X_0 = (-0.2, 0, 0)$.

In this experiment we put 5% of gaussian noise in (f, Φ). The obtained results are illustrated in figure-1, where the reconstructed fields u_1 and u_2 are presented at convergence. The non radial missing data are presented in figure-2 and relative L^2-errors on the reconstructed fields and boundary data are shown in figure-3.

2.2. *Non-homogeneous test-case*

The domain Ω is the open bounded set delimited by two concentric spheres in 3D, of radius $R_c = 1$ and $R_i = 0.6$. The inner boundary plays the role of Γ_i, where the trace and normal derivative are missing, and the outer one plays the one of Γ_c where the latter information is over specified. However, a third sphere of radius $R_m = 0.8$ is added between Γ_i and Γ_c.

We consider a non-homogeneous conductivity k : equal to 1 in the domain between R_c and R_m, and is an orthotropic tensor field $I_2 \otimes (\alpha I_1)$ in the domain delimited by R_m and R_i ($\alpha = 0.5$).

The Cauchy data are generated via a finite element solution of the mixed boundary value problem $\nabla.k\nabla u = 0$ in Ω, $k\nabla u.\nu_{|\Gamma_c} = \Phi = 0$, and $u_{|\Gamma_i} = \frac{1}{\sqrt{(x+0.2)^2+y^2+z^2}}$.

The data and reference solution are represented in figure-4. After convergence of the Nash iterations, we obtain a relative error less than 1.5% on the reconstructed fields and on the missing data, see figure-5 where the Cauchy data are perturbed with a Gaussian (5% level) noise. In figure-6, we illustrate how the Nash iterations yield, as a byproduct, a regularized noise-free approximation of noisy Cauchy data.

From a computational point of view, let us underline an interesting benefit of using a game approach, namely the possibility of using, independently, optimization algorithms of very different kinds -if necessary- to solve the respective players minimization subproblems in (8).

3. Conclusion

Let us conclude the paper with some short remarks. First of all, we have used the simplest class of games to model the completion problem, namely the class of static games with complete information. This simple game formulation yields interesting results like the existence and uniqueness of a Nash equilibrium even when the Cauchy data are not compatible, and also the fact that the Nash equilibrium is the missing data when the Cauchy problem has a solution.

Investigation of more sophisticated classes of games such as dynamical games with incomplete information may lead to new efficient data completion algorithms. It is also interesting to notice that solving the data completion problem with our method makes use of the standard computational tools, be it finite element or optimization codes. The numerical experiments presented for different test-cases prove that our method exhibits remarkable numerical stability with respect to noisy Cauchy data, and numerous additional 2D computational experiments (not presented here) do corroborate the efficiency and stability of the presented algorithm to tackle the Cauchy problem.

References

[1] Aboulaïch R, Ben Abda A and Kallel M 2008 Missing boundary data reconstruction via an approximate optimal control *Inverse Problems and Imaging* **2** 411–426
[2] Alessandrini G, Rondi L, Rosset E and Vessella S 2009 The stability for the Cauchy problem for elliptic equations *Inverse Problems* **25** 123004 (47pp)
[3] Attouch H, Bolte J, Redont P and Soubeyran A 2008 Alternating proximal algorithms for weakly coupled convex minimization problems. Applications to dynamical games and PDE's *J. Convex Anal.* **15** 485–506
[4] Hadamard J 1953 *Lectures on Cauchy's Problem in Linear Partial Differential Equation*, Dover, New york USA
[5] Kozlov V A, Maz'ya V G and Fomin A V 1991 An iterative method for solving the Cauchy problems for elliptic equations *Comput. Math. Phys.* **31** 45–52
[6] Li S and Basar T 1987 Distributed algorithms for the computation of noncooperative equilibria *Automatica* **23** 523–533

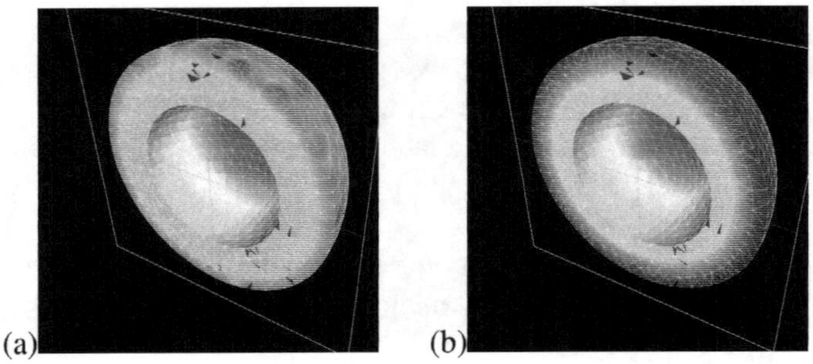

Figure 1. Non radial case. Noise level is 5%. The level sets of the reconstructed fields (a) u_1 and (b) u_2 are presented at convergence. The Finite Element computations are performed with 4740 nodes and 22795 tetrahedral elements.

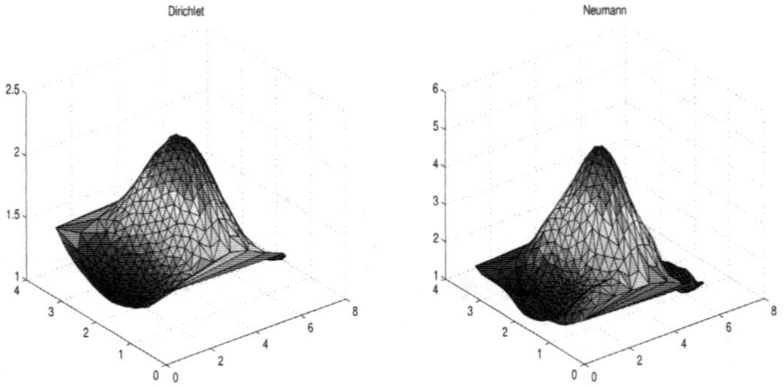

Figure 2. Non radial case. Reconstructed non radial Dirichlet (τ_N, left) and Neumann (η_N, right) data over Γ_i. The profiles are presented at convergence and for a noise level $\sigma = 5\%$.

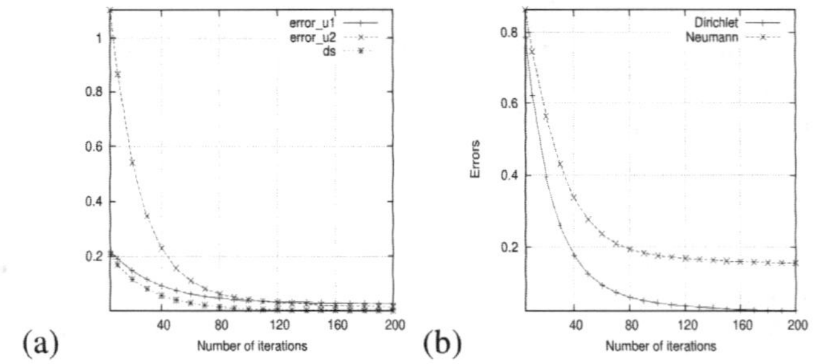

Figure 3. Non radial case. Relative L^2-errors are presented for the non radial case as a function of the overall Nash iterations. Noise level is 5%. (a) reconstructed fields error : $\|u_i^{(k)} - u\|/\|u\|$, $i = 1, 2$ and Nash strategies $ds = \|S^{(k)} - S^{(k-1)}\|$; (b) missing Dirichlet data : $\|\tau^{(k)} - u_{|\Gamma_i}\|/\|u_{|\Gamma_i}\|$ and Neumann data $\|\eta^{(k)} - \frac{\partial u}{\partial \nu}_{|\Gamma_i}\|/\|\frac{\partial u}{\partial \nu}_{|\Gamma_i}\|$.

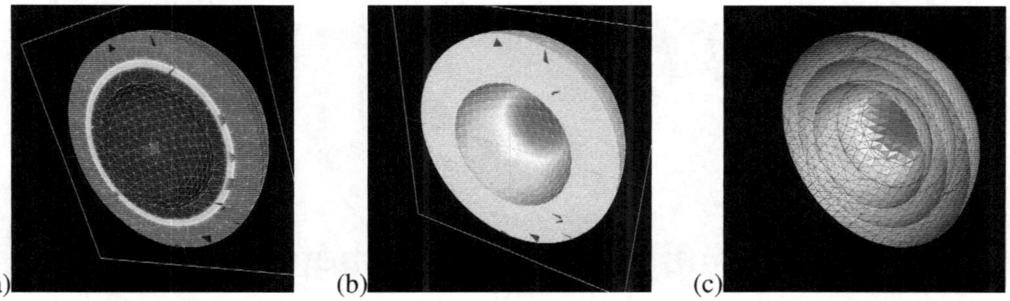

(a) (b) (c)

Figure 4. Orthotropic case. Data (a) conductivity distribution (b) The reference solution u level sets in Ω and (c) Traces of u on the three interfaces Γ_c, Γ_m and Γ_i. The FE mesh has 4560 nodes and 21685 tetrahedral elements

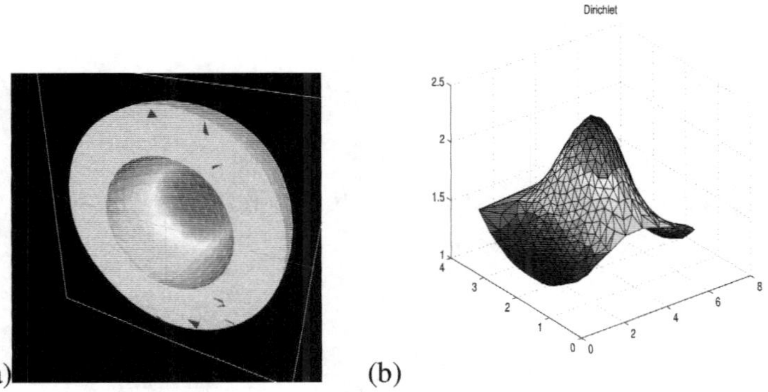

(a) (b)

Figure 5. Orthotropic case. Reconstruction. The solutions are presented at convergence and for a noise level $\sigma = 5\%$. (a) reconstructed fields u_2 with: $\|u_2 - u_1\|/\|u\| = 0.013$ and $\|u_2 - u\|/\|u\| = 0.014$; (b) missing Dirichlet data : $\|\tau_N - u_{|\Gamma_i}\|/\|u_{|\Gamma_i}\| = 0.015$.

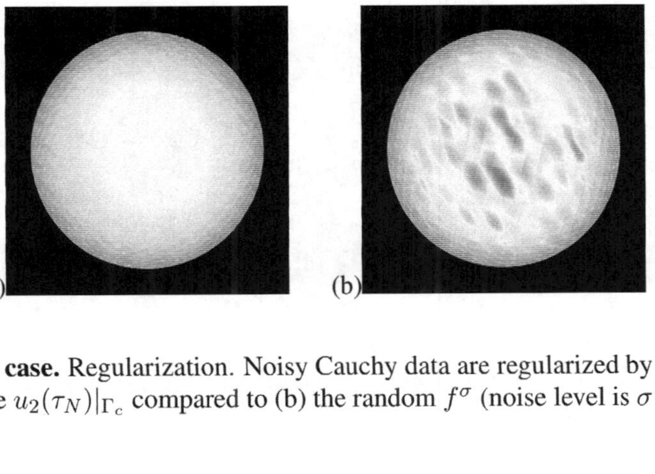

(a) (b)

Figure 6. Orthotropic case. Regularization. Noisy Cauchy data are regularized by the Nash process. (a) the smoothed profile $u_2(\tau_N)|_{\Gamma_c}$ compared to (b) the random f^σ (noise level is $\sigma = 5\%$).

Constrained variable projection method for blind deconvolution

A Cornelio[1], E Loli Piccolomini[2], J G Nagy[3]

[1]Department of Pure and Applied Mathematics, University of Modena and Reggio Emilia
[2]Department of Mathematics, University of Bologna
[3]Department of Mathematics and Computer Science, Emory University

E-mail: anastasia.cornelio@unimore.it

Abstract. This paper is focused on the solution of the blind deconvolution problem, here modeled as a separable nonlinear least squares problem. The well known ill-posedness, both on recovering the blurring operator and the true image, makes the problem really difficult to handle. We show that, by imposing appropriate constraints on the variables and with well chosen regularization parameters, it is possible to obtain an objective function that is fairly well behaved. Hence, the resulting nonlinear minimization problem can be effectively solved by classical methods, such as the Gauss-Newton algorithm.

1. Introduction

Blind deconvolution can be modeled as an inverse problem of the form [1]

$$\boldsymbol{b} = \boldsymbol{A}(\boldsymbol{y}_{\text{true}})\boldsymbol{x}_{\text{true}} + \boldsymbol{\eta} \tag{1}$$

where $\boldsymbol{b} \in \mathcal{R}^m$ is the measured, blurred and noisy image ($\boldsymbol{\eta}$ models the noise), and $\boldsymbol{x}_{\text{true}} \in \mathcal{R}^n$ represents the unknown true image. The vector $\boldsymbol{y}_{\text{true}} \in \mathcal{R}^p$ is unknown and $\boldsymbol{A}(\boldsymbol{y})$ is a nonlinear operator, which models the blurring and maps $\boldsymbol{y}$ into an $m \times n$ matrix. We call equation (1) a *separable* inverse problem because the measured data depends linearly on the unknown vector $\boldsymbol{x}$ and nonlinearly on the unknown vector $\boldsymbol{y}$.

In [1] the authors, once defined the model mapping $\boldsymbol{A}(\cdot)$ and given the noisy data $\boldsymbol{b}$, formulate the optimization problem

$$\min_{\boldsymbol{x},\boldsymbol{y}} f_0(\boldsymbol{x}, \boldsymbol{y}) \tag{2}$$

and then propose to apply the variable projection method [2, 3, 4, 5] in order to determine the unknowns $\boldsymbol{x}$ and $\boldsymbol{y}$. By taking advantage of the separability of the variables, the method consists on eliminating the linear variables $\boldsymbol{x}$ thus obtaining a reduced nonlinear cost functional, that depends only on $\boldsymbol{y}$:

$$f(\boldsymbol{y}) = f_0(\boldsymbol{x}(\boldsymbol{y}), \boldsymbol{y}). \tag{3}$$

Aim of this work is to show that imposing appropriate constraints on the unknowns provides a better behaved objective function, which results in computing a better solution of problem (2) when using a standard optimization method such as Gauss-Newton. In particular, we imposed nonnegative constraints on the linear variables $\boldsymbol{x}$ and stated the problem as a least squares problems with Tikhonov regularization on $\boldsymbol{x}$, that is

$$\min_{\boldsymbol{x},\boldsymbol{y}} f_0(\boldsymbol{x},\boldsymbol{y}) = \|\boldsymbol{A}(\boldsymbol{y})\boldsymbol{x} - \boldsymbol{b}\|_2^2 + \lambda\|\boldsymbol{x}\|_2^2 \quad \text{subject to } \boldsymbol{x} \geq \boldsymbol{0}. \tag{4}$$

Equation (2) is still well defined if linear constraints are imposed on $\boldsymbol{x}$. In [5] Sima and Van Huffel exploit this property and provide results for an application in spectroscopy. Since the problem is well conditioned in this framework, regularization is not needed. The paper is organized as follows. In Section 2 the variable projection method is extended to the nonnegatively constrained case; in Section 3 we investigate the behaviour of the objective function in the presence of constraints, while in Section 4 we provide some experiments. Finally, conclusions are given in Section 5.

2. Variable Projection for nonnegative constraints

The variable projection method [2, 3, 5] has been applied to blind deconvolution modeled as a separable nonlinear least squares problem in [1]. In the following, we will extend the method to the case of nonnegative constraints imposed on $\boldsymbol{x}$.

If we knew the nonlinear parameters $\boldsymbol{y}$, then the linear variables $\boldsymbol{x}(\boldsymbol{y})$ could be obtained by solving the *nonnegative* linear least-squares problem:

$$\boldsymbol{x}(\boldsymbol{y}) = \min_{\boldsymbol{x}\geq\boldsymbol{0}} \|\boldsymbol{A}(\boldsymbol{y})\boldsymbol{x} - \boldsymbol{b}\|_2^2 + \lambda\|\boldsymbol{x}\|_2^2. \tag{5}$$

By replacing (5) in the original objective function (4), we obtain a new reduced cost function, depending only on $\boldsymbol{y}$

$$\min_{\boldsymbol{y}} f(\boldsymbol{y}) = f_0(\boldsymbol{x}(\boldsymbol{y}),\boldsymbol{y}). \tag{6}$$

Now, like in any nonnegatively constrained optimization problem, for any $\boldsymbol{x} \geq 0$ we can define: the active set by $\mathcal{A}(\boldsymbol{x}) = \{i \,|\, x_i = 0\}$ and a diagonal matrix $\boldsymbol{D}(\boldsymbol{x})$ by

$$\boldsymbol{D}(\boldsymbol{x})_{ii} = \begin{cases} 1 & \text{if} \quad i \notin \mathcal{A}(\boldsymbol{x}) \\ 0 & \text{if} \quad i \in \mathcal{A}(\boldsymbol{x}) \end{cases} .$$

In the following we will use the notation $\boldsymbol{D} = \boldsymbol{D}(\boldsymbol{x}(\boldsymbol{y}))$, where $\boldsymbol{x}(\boldsymbol{y})$ is defined by (5). Therefore the solution of (5) can be written by a closed formula as in [6]:

$$\boldsymbol{x}(\boldsymbol{y}) = \begin{bmatrix} \boldsymbol{A}(\boldsymbol{y})\boldsymbol{D} \\ \lambda \boldsymbol{I}_n \end{bmatrix}^{\dagger} \begin{bmatrix} \boldsymbol{b} \\ \boldsymbol{0} \end{bmatrix} \tag{7}$$

where $\dagger$ indicates the pseudoinverse of a matrix and $\boldsymbol{0}$ is the n-dimensional zero column vector. By using (7), (6) becomes exactly

$$\min_{\boldsymbol{y}} f(\boldsymbol{y}) = \left\| \boldsymbol{A}(\boldsymbol{y}) \begin{bmatrix} \boldsymbol{A}(\boldsymbol{y})\boldsymbol{D} \\ \lambda \boldsymbol{I}_n \end{bmatrix}^{\dagger} \begin{bmatrix} \boldsymbol{b} \\ \boldsymbol{0} \end{bmatrix} - \boldsymbol{b} \right\|_2^2 \tag{8}$$

which can be solved with a classical nonlinear least squares method, such as the Gauss-Newton algorithm. The computation of the Jacobian $\boldsymbol{J}$ of $f(\boldsymbol{y})$ is widely discussed in [7].

3. Role of the constraints

In this section we study the behavior of the objective function $f(\boldsymbol{y})$ defined in (8), in order to investigate the effects of the constraints on the solution of the outer nonlinear problem. To simplify the representation of the function, we consider $\boldsymbol{y}$ as containing just one parameter. In particular, in the following tests we consider the 2-dimensional Gaussian PSF defined as

$$p_{ij} = \exp\left(-\frac{1}{2}\begin{bmatrix} i-k \\ j-l \end{bmatrix}^T \begin{bmatrix} \sigma & 0 \\ 0 & \sigma \end{bmatrix}^{-1} \begin{bmatrix} i-k \\ j-l \end{bmatrix}\right) \tag{9}$$

where $\boldsymbol{y} = (\sigma, \sigma)$, σ determines the spread of the PSF in both the directions and (k, l) is the center of the PSF.

The goal of this test is to show that the nonnegative constraints imposed on $\boldsymbol{x}$ have a crucial role in our application. In the following, we use the $\boldsymbol{x}$ is the vectorized 256×256 satellite image (Figure 2 (b)), blurred by the Gaussian PSF (9) with $\sigma = 2.5$ and corrupted by 5% Gaussian white noise.

We compare the function $f(\boldsymbol{y})$ in (8) with the function $g(\boldsymbol{y})$ obtained without imposing constraints on $\boldsymbol{x}$, i.e.:

$$g(\boldsymbol{y}) = \left\| \boldsymbol{A}(\boldsymbol{y}) \begin{bmatrix} \boldsymbol{A}(\boldsymbol{y}) \\ \lambda \boldsymbol{I}_n \end{bmatrix}^\dagger \begin{bmatrix} \boldsymbol{b} \\ \boldsymbol{0} \end{bmatrix} - \boldsymbol{b} \right\|_2^2, \tag{10}$$

In Figure 1 we compare the objective functions $g(\boldsymbol{y})$ and $f(\boldsymbol{y})$ in this easier 1-dimensional case. Let us start with the unconstrained formulation of the problem (Figure 1 (a)). We solve the inner linear problem in (10) with the unconstrained hybrid (HyBR) method proposed in [1]. By using the regularization parameter λ automatically chosen by HyBR, we obtain that $g(\boldsymbol{y})$ is a monotonically increasing function (black dashdot line). Then we use the optimal λ, obtained when the true $\boldsymbol{x}$ solution is given: now the function is convex (red plain line), but has an unique global minimum that does not match the true value $\boldsymbol{y} = 2.5$.

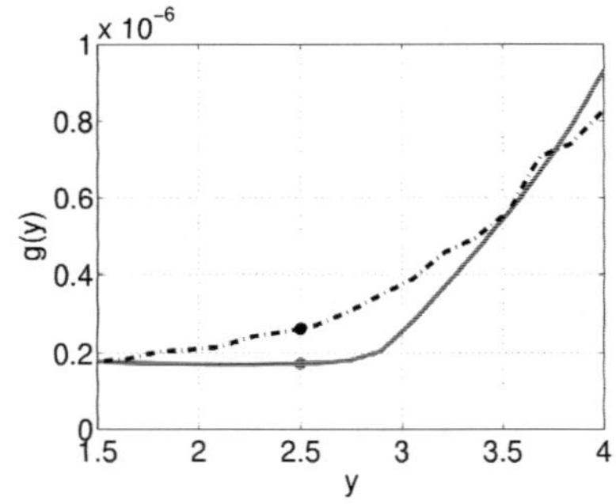

(a) $g(\boldsymbol{y})$

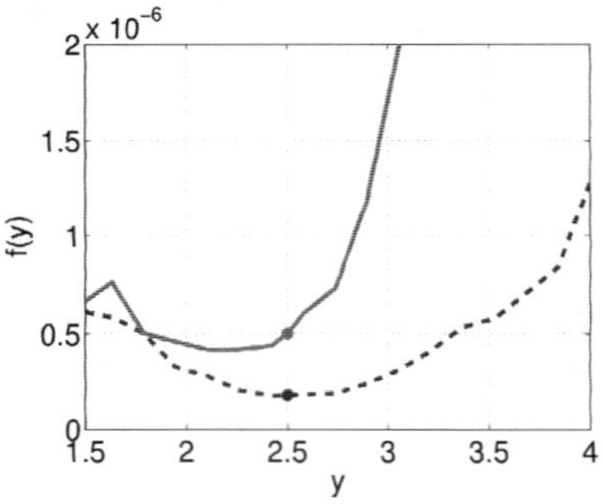

(b) $f(\boldsymbol{y})$

Figure 1. Left is the plot of $g(\boldsymbol{y})$ with λ_{HyBR} automatically chosen (black dashdot line) and with the optimal λ_{OPT} (red plain line). Right is the plot of $f(\boldsymbol{y})$: with λ_{HyBR} (red plain line) and with a smaller value equal to $\lambda_{HyBR}/200$ (blue dashed line).

Let us consider now the nonnegative function $f(\boldsymbol{y})$ (Figure 1 (b)), where the inner linear problem has been solved by a scaled gradient projection (SGP) method [8]. Since we are not provided by a rule to choose the regularization parameter in the presence of nonnegative constraints, we use the optimal λ obtained by the previous unconstrained method (red plain line). Let us remind that imposing nonnegative constraints actually reduces the ill-conditioning of the problem [6]. For this reason it makes sense using, in this case, a smaller λ than in the unconstrained problem: the global minimum matches the true solution (blue dashed line).

Journal of Physics: Conference Series **386** (2012) 012005 doi:10.1088/1742-6596/386/1/012005

4. Numerical results

In this section we show how imposing nonnegative constraints on $\boldsymbol{x}$ provides better results for both the linear and nonlinear variables. In particular we consider as true objects the vectorized 256×256 galaxy and the satellite images (Figure 2), blurred by the Moffat PSF

$$p_{ij} = \left(1 + \begin{bmatrix} i-k \\ j-l \end{bmatrix}^T \begin{bmatrix} \sigma_1 & 0 \\ 0 & \sigma_2 \end{bmatrix}^{-1} \begin{bmatrix} i-k \\ j-l \end{bmatrix}\right)^{-\beta}$$

defined by the parameter $\boldsymbol{y}_{true} = (\sigma_1, \sigma_2) = (1, 2)$ and corrupted by different levels of white

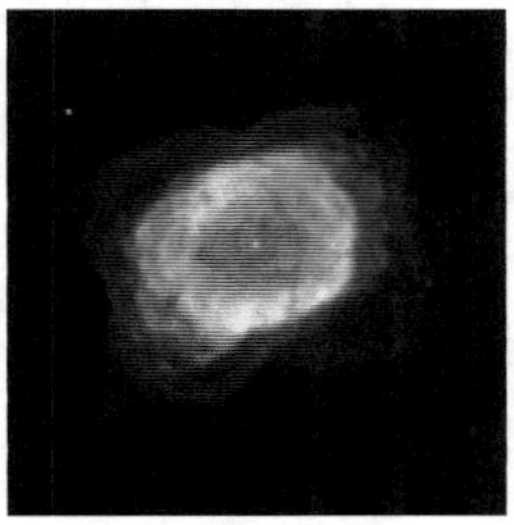

(a) galaxy

(b) satellite

Figure 2. True objects

Gaussian noise. The regularization parameter λ has been fixed for all the Gauss-Newton iterations and it has been heuristically chosen as the parameter providing the best results in terms of relative error. In Table 1 we compare the unconstrained (10) and the constrained (8) formulation of the blind deconvolution problem. As indicators of the reconstruction quality, the minimum of the $\boldsymbol{x}$ and $\boldsymbol{y}$ relative errors and the structural similarity index [9] (SSIM) are shown.

Table 1. Results of galaxy image

	3% noise			5% noise		
problem	err_x	err_y	SSIM	err_x	err_y	SSIM
Unconstrained: $\min g$	0.197159	0.225329	0.44505	0.218703	0.196787	0.37440
Constrained: $\min f$	0.072446	0.031953	0.83498	0.076305	0.033387	0.81646

Moreover, in Figure 3 we consider the satellite image corrupted by 3% noise and blurred with $\boldsymbol{y} = (1.5, 2)$; we compare the relative errors obtained by solving the unconstrained problem (10) (black triangles) and of the constrained problem (8) (magenta squares). Figure 3 (c) confirms that the function $g(\boldsymbol{y})$ is still decreasing, but the minimum we're going to reach is far from the auspicable solution of the problem. On the other hand, $f(\boldsymbol{y})$ becomes flat during the iterations: this suggests a stopping criterion for the Gauss-Newton algorithm, that involves the objective function, such as

$$\frac{|f(\boldsymbol{x}_{k+1}) - f(\boldsymbol{x}_k)|}{f(\boldsymbol{x}_{k+1})} \leq \frac{\eta}{2}.$$

We used as a good value for η the level of noise in the observed data.

Journal of Physics: Conference Series **386** (2012) 012005 doi:10.1088/1742-6596/386/1/012005

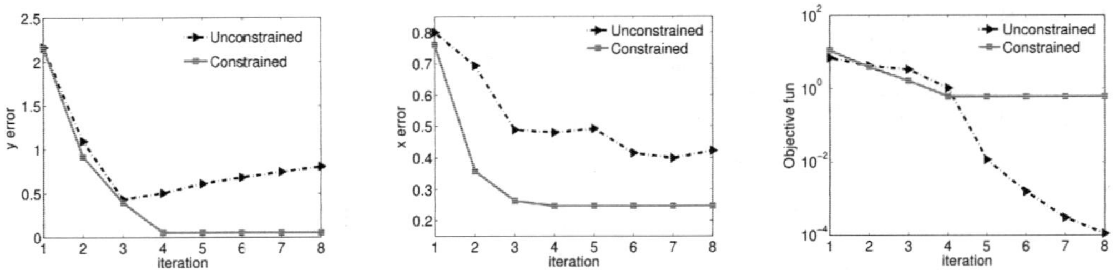

(a) $\boldsymbol{y}$ relative errors (b) $\boldsymbol{x}$ relative errors (c) Objective functions

Figure 3. Relative errors of y (left) and x (middle). Right is the objective function during the Gauss-Newton iterations.

5. Conclusions

In this paper we have illustrated that a substantial amount of uncertainty arises when solving blind deconvolution problems, firstly caused not only by noise and measurement errors, but also by imprecise regularization parameters and linear solvers. We showed that imposing constraints on some parameters, i.e. solving the ill-posed problem with greater accuracy, produces an objective function more suitable for the common optimization algorithms, such as Gauss-Newton. This suggests that a priori knowledge about the application is useful to include appropriate constraints on $\boldsymbol{x}$ and $\boldsymbol{y}$. However, to efficiently solve this kind of separable ill-posed inverse problems, a substantial amount of work is needed in order to have a robust method. First of all, the development of methods to automatically choose regularization parameters for constrained deconvolution problems could improve significantly the solution of this kind of nonlinear inverse problems.

References

[1] Chung J and Nagy J 2010 *SIAM J. Sci. Comput.* **31** 4654–4674
[2] Golub G H and Pereyra V 1973 *SIAM J. Numer. Anal.* **10** 413–432
[3] Kaufman L 1975 *BIT* **15** 49–57
[4] Kaufman L and Pereyra V 1978 *SIAM J. Numer. Anal.* **15** 12–20
[5] Sima D M and Huffel S V 2007 *J. Comput. Appl. Math.* **203** 264–278
[6] Bardsley J M, Merikoski J and Vio R 2008 *IJPAM* **43** 95–109
[7] Cornelio A, Piccolomini E L and Nagy J G 2012 Numerical optimization methods for blind deconvolution Tech. rep. URL `http://www.optimization-online.org/DB_HTML/2012/02/3351.html`
[8] Bonettini S, Zanella R and Zanni L 2009 *Inverse Problems* **25**
[9] Wang Z, Bovik A, Sheikh H and Simoncelli E 2004 *IEEE Trans. Image Proc.* **13** 600–612

2nd Int. Workshop on New Computational Methods for Inverse Problems (NCMIP 2012) IOP Publishing
Journal of Physics: Conference Series **386** (2012) 012006 doi:10.1088/1742-6596/386/1/012006

Variational Bayesian Approximation methods for inverse problems

Ali Mohammad-Djafari

Laboratoire des signaux et systèmes (L2S),
UMR 8506 CNRS-SUPELEC-UNIV PARIS SUD,
SUPELEC, Plateau de Moulon, 91192 Gif-sur-Yvette, France

E-mail: djafari@lss.supelec.fr

Abstract. Variational Bayesian Approximation (VBA) methods are recent tools for effective Bayesian computations. In this paper, these tools are used for inverse problems where the prior models include hidden variables and where where the estimation of the hyper parameters has also to be addressed. In particular two specific prior models (Student-t and mixture of Gaussian models) are considered and details of the algorithms are given.

1. Introduction

In many generic inverse problems in signal and image processing, the problem is to infer on an unknown signal $f(t)$ or an unknown image $f(\boldsymbol{r})$ with $\boldsymbol{r} = (x, y)$ through an observed signal $g(t')$ or an observed image $g(\boldsymbol{r}')$ related between them through an operator $\mathcal{H}$ such as convolution $g = h * f$ or any other linear or non linear transformation $g = \mathcal{H}f$. When this relation is linear and we have discretized the problem, we arrive to the relation: $\boldsymbol{g} = \boldsymbol{H}\boldsymbol{f} + \boldsymbol{\epsilon}$, where $\boldsymbol{f} = [f_1, \cdots, f_n]'$ represents the unknowns, $\boldsymbol{g} = [g_1, \cdots, g_m]'$ the observed data, $\boldsymbol{\epsilon} = [\epsilon_1, \cdots, \epsilon_m]'$ the errors of modelling and measurement and $\boldsymbol{H}$ the matrix of the system response.

The Bayesian inference approach is based on the posterior law:

$$p(\boldsymbol{f}|\boldsymbol{g}, \boldsymbol{\theta}_1, \boldsymbol{\theta}_2) = \frac{p(\boldsymbol{g}|\boldsymbol{f}, \boldsymbol{\theta}_1)\, p(\boldsymbol{f}|\boldsymbol{\theta}_2)}{p(\boldsymbol{g}|\boldsymbol{\theta}_1, \boldsymbol{\theta}_2)} \propto p(\boldsymbol{g}|\boldsymbol{f}, \boldsymbol{\theta}_1)\, p(\boldsymbol{f}|\boldsymbol{\theta}_2) \tag{1}$$

where the sign $\propto$ stands for "proportional to", $p(\boldsymbol{g}|\boldsymbol{f}, \boldsymbol{\theta}_1)$ is the likelihood, $p(\boldsymbol{f}|\boldsymbol{\theta}_2)$ the prior model, $\boldsymbol{\theta} = (\boldsymbol{\theta}_1, \boldsymbol{\theta}_2)$ are their corresponding parameters (often called the hyper-parameters of the problem) and $p(\boldsymbol{g}|\boldsymbol{\theta}_1, \boldsymbol{\theta}_2)$ is called the evidence of the model. When the parameters $\boldsymbol{\theta}$ have to be estimated too, a prior $p(\theta|\boldsymbol{\theta}_0)$ with fixed values for $\boldsymbol{\theta}_0$ is assigned to them and the expression of the joint posterior

$$p(\boldsymbol{f}, \theta|\boldsymbol{g}, \boldsymbol{\theta}_0) = \frac{p(\boldsymbol{g}|\boldsymbol{f}, \theta_1)\, p(\boldsymbol{f}|\theta_2)\, p(\theta|\boldsymbol{\theta}_0)}{p(\boldsymbol{g}|\boldsymbol{\theta}_0)} \tag{2}$$

is used to infer them jointly.

Variational Bayesian Approximation (BVA) methods try to approximate $p(\boldsymbol{f}, \theta|\boldsymbol{g})$ by a separable one $q(\boldsymbol{f}, \theta|\boldsymbol{g}) = q_1(\boldsymbol{f}|\widetilde{\boldsymbol{\theta}}, \boldsymbol{g})\; q_2(\theta|\widetilde{\boldsymbol{f}}, \boldsymbol{g})$ and then using them for estimation [1, 2, 3, 4, 5, 6, 7, 8, 9].

For hierarchical prior models with hidden variables z, the problem becomes more complex, because we have to give the expression of the joint posterior law

$$p(f,z,\theta|\boldsymbol{g}) \propto p(\boldsymbol{g}|f,\theta_1)\,p(f|z,\theta_2)\,p(z|\theta_3)\,p(\theta|\boldsymbol{\theta}_0) \tag{3}$$

and then approximate it by a separable one

$$q(f,z,\theta|\boldsymbol{g}) = q_1(f|\widetilde{\boldsymbol{z}},\widetilde{\boldsymbol{\theta}},\boldsymbol{g})\;q_2(z|\widetilde{\boldsymbol{f}},\widetilde{\boldsymbol{\theta}},\boldsymbol{g})\;q_3(\theta|\widetilde{\boldsymbol{z}},\widetilde{\boldsymbol{f}},\boldsymbol{g}) \tag{4}$$

and then using them for estimation.

In this paper, first the general VBA method is detailed for the inference on inverse problems with hierarchical prior models. Then, two particular classes of prior models (Student-t and mixture of Gaussians) are considered and the details of BVA algorithms for them are given.

2. Bayesian Variational Approximation with hierarchical prior models

When a hierarchical prior model $p(f|z,\theta)$ is used and when the estimation of the hyper-parameters $\boldsymbol{\theta}$ has to be considered, the joint posterior law of all the unknowns becomes:

$$p(f,z,\theta|\boldsymbol{g}) \propto p(\boldsymbol{g}|f,\theta_1)\,p(f|z,\theta_2)\,p(z|\theta_3)\,p(\theta) \tag{5}$$

The main idea behind the VBA is to approximate the joint posterior $p(f,z,\theta|\boldsymbol{g})$ by a separable one, for example

$$q(f,z,\theta|\boldsymbol{g}) = q_1(f|\boldsymbol{g})\,q_2(z|\boldsymbol{g})\,q_3(\theta|\boldsymbol{g}) \tag{6}$$

and where the expressions of $q(f,z,\theta|\boldsymbol{g})$ is obtained by minimizing the Kullback-Leibler divergence

$$\mathrm{KL}(q:p) = \int q \ln\frac{q}{p} = \left\langle \ln\frac{q}{p}\right\rangle_q \tag{7}$$

It is then easy to show that $\mathrm{KL}(q:p) = \ln p(\boldsymbol{g}|\mathcal{M}) - \mathcal{F}(q)$ where $p(\boldsymbol{g}|\mathcal{M})$ is the likelihood of the model

$$p(\boldsymbol{g}|\mathcal{M}) = \int\!\!\int\!\!\int p(f,z,\theta,\boldsymbol{g}|\mathcal{M})\;\mathrm{d}f\;\mathrm{d}z\;\mathrm{d}\theta \tag{8}$$

with $p(f,z,\theta,\boldsymbol{g}|\mathcal{M}) = p(\boldsymbol{g}|f,\theta)\,p(f|z,\theta)\,p(z|\theta)\,p(\theta)$ and $\mathcal{F}(q)$ is the free energy associated to q defined as

$$\mathcal{F}(q) = \left\langle \ln\frac{p(f,z,\theta,\boldsymbol{g}|\mathcal{M})}{q(f,z,\theta)}\right\rangle_q \tag{9}$$

So, for a given model $\mathcal{M}$, minimizing $\mathrm{KL}(q:p)$ is equivalent to maximizing $\mathcal{F}(q)$ and when optimized, $\mathcal{F}(q^*)$ gives a lower bound for $\ln p(\boldsymbol{g}|\mathcal{M})$.

Without any other constraint than the normalization of q, an alternate optimization of $\mathcal{F}(q)$ with respect to q_1, q_2 and q_3 results in

$$\left\{\begin{array}{l} q_1(f) \propto \exp\left\{-\left\langle \ln p(f,z,\theta,\boldsymbol{g})\right\rangle_{q(z)q(\theta)}\right\}, \\ q_2(z) \propto \exp\left\{-\left\langle \ln p(f,z,\theta,\boldsymbol{g})\right\rangle_{q(f)q(\theta)}\right\} \\ q_3(\theta) \propto \exp\left\{-\left\langle \ln p(f,z,\theta,\boldsymbol{g})\right\rangle_{q(f)q(z)}\right\} \end{array}\right. \tag{10}$$

Note that these relations represent an implicit solution for $q_1(f)$, $q_2(z)$ and $q_3(\theta)$ which need, at each iteration, the expression of the expectations in the right hand of exponentials. If $p(\boldsymbol{g}|f,z,\theta_1)$ is a member of an exponential family and if all the priors $p(f|z,\theta_2)$, $p(z|\theta_3)$, $p(\theta_1)$, $p(\theta_2)$, and $p(\theta_3)$ are conjugate priors, then it is easy to see that these expressions leads to standard distributions for which the required expectations are easily evaluated. In that case, we may note

$$q(f,z,\theta|\boldsymbol{g}) = q_1(f|\widetilde{\boldsymbol{z}},\widetilde{\boldsymbol{\theta}};\boldsymbol{g})\, q_2(z|\widetilde{\boldsymbol{f}},\widetilde{\boldsymbol{\theta}};\boldsymbol{g})\, q_3(\theta|\widetilde{\boldsymbol{f}},\widetilde{\boldsymbol{z}};\boldsymbol{g}) \tag{11}$$

where the tilded quantities $\widetilde{\boldsymbol{z}}$, $\widetilde{\boldsymbol{f}}$ and $\widetilde{\boldsymbol{\theta}}$ are, respectively functions of $(\widetilde{\boldsymbol{f}},\widetilde{\boldsymbol{\theta}})$, $(\widetilde{\boldsymbol{z}},\widetilde{\boldsymbol{\theta}})$ and $(\widetilde{\boldsymbol{f}},\widetilde{\boldsymbol{z}})$ and where the alternate optimization results to alternate updating of the parameters $(\widetilde{\boldsymbol{z}},\widetilde{\boldsymbol{\theta}})$ for q_1, the parameters $(\widetilde{\boldsymbol{f}},\widetilde{\boldsymbol{\theta}})$ of q_2 and the parameters $(\widetilde{\boldsymbol{f}},\widetilde{\boldsymbol{z}})$ of q_3. Finally, we may note that, to monitor the convergence of the algorithm, we may evaluate the free energy

$$\begin{aligned}\mathcal{F}(q) = \langle \ln p(f,z,\theta,\boldsymbol{g}|\mathcal{M})\rangle_q + \langle -\ln q(f,z,\theta)\rangle_q &= \langle \ln p(\boldsymbol{g}|f,z,\theta)\rangle_q + \langle \ln p(f|z,\theta)\rangle_q + \langle \ln p(z|\theta)\rangle_q \\ &+ \langle -\ln q(f)\rangle_q + \langle -\ln q(z)\rangle_q + \langle -\ln q(\theta)\rangle_q\end{aligned} \tag{12}$$

where all the expectations are with respect to q. Other decompositions are also possible:

$$q(f,z,\theta|\boldsymbol{g}) = \prod_j q_{1j}(f_j|\widetilde{\boldsymbol{f}}_{(-j)},\widetilde{\boldsymbol{z}},\widetilde{\boldsymbol{\theta}};\boldsymbol{g}) \prod_j q_{2j}(z_j|\widetilde{\boldsymbol{f}},\widetilde{\boldsymbol{z}}(-j),\widetilde{\boldsymbol{\theta}};\boldsymbol{g}) \prod_l q_{3l}(\theta_l|\widetilde{\boldsymbol{f}},\widetilde{\boldsymbol{z}},\widetilde{\boldsymbol{\theta}}_{(-l)}\ \boldsymbol{g}) \tag{13}$$

or

$$q(f,z,\theta|\boldsymbol{g}) = q_1(f|\widetilde{\boldsymbol{z}},\widetilde{\boldsymbol{\theta}};\boldsymbol{g}) \prod_j q_{2j}(z_j|\widetilde{\boldsymbol{f}},\widetilde{\boldsymbol{z}}_{(-j)},\widetilde{\boldsymbol{\theta}};\boldsymbol{g}) \prod_l q_{3l}(\theta_l|\widetilde{\boldsymbol{f}},\widetilde{\boldsymbol{z}},\widetilde{\boldsymbol{\theta}}_{(-l)};\boldsymbol{g}) \tag{14}$$

Here, we consider this case and give some more details on it.

3. Bayesian Variational Approximation with Student-t priors

The Student-t model is:

$$p(f|\nu) = \prod_j \mathcal{S}t(f_j|\nu) \text{ with } \mathcal{S}t(f_j|\nu) = \frac{1}{\sqrt{\pi\nu}} \frac{\Gamma((\nu+1)/2)}{\Gamma(\nu/2)} \left(1 + f_j^2/\nu\right)^{-(\nu+1)/2} \tag{15}$$

Knowing that

$$\mathcal{S}t(f_j|\nu) = \int_0^\infty \mathcal{N}(f_j|0,1/\tau_j)\, \mathcal{G}(\tau_j|\nu/2,\nu/2)\, \mathrm{d}\tau_j \tag{16}$$

we can write this model via the positive hidden variables τ_j:

$$\begin{aligned} p(f|\tau) &= \prod_j p(f_j|\tau_j) = \prod_j \mathcal{N}(f_j|0,1/\tau_j) \propto \exp\left\{-\tfrac{1}{2}\sum_j \tau_j f_j^2\right\} \\ p(\tau_j|\alpha,\beta) &= \mathcal{G}(\tau_j|\alpha,\beta) \propto \tau_j^{(\alpha-1)} \exp\{-\beta\tau_j\} \text{ with } \alpha=\beta=\nu/2 \end{aligned} \tag{17}$$

Cauchy model is obtained when $\nu = 1$:

In this case, let consider the forward model $\boldsymbol{g} = \boldsymbol{H}\boldsymbol{f} + \boldsymbol{\epsilon}$ and assign a Gaussian law to the noise $\boldsymbol{\epsilon}$ which which results to $p(\boldsymbol{g}|\boldsymbol{f},v_\epsilon) = \mathcal{N}(\boldsymbol{g}|\boldsymbol{H}\boldsymbol{f},v_\epsilon\boldsymbol{I})$. We also assign a prior $p(\tau_\epsilon|\alpha_{\tau0},\beta_{\tau0}) = \mathcal{G}(\tau_\epsilon|\alpha_{\tau0},\beta_{\tau0})$ to $\tau_\epsilon = 1/v_\epsilon$. Let also note $\boldsymbol{\tau} = [\tau_1,\cdots,\tau_N]$, $\boldsymbol{T} = \mathrm{diag}\,[\tau]$, $z_j = 1/\tau_j$, $Z = \mathrm{diag}\,[z] = \boldsymbol{T}^{-1}$ and note $p(f|\tau) = \prod_j p(f_j|\tau_j) = \prod_j \mathcal{N}(f_j|0,\tau_j) = \mathcal{N}(f|0,\boldsymbol{T})$ and finally, assign $p(\tau|\alpha_0,\beta_0) = \prod_j \mathcal{G}(\tau_j|\alpha_0,\beta_0)$.

Then, we obtain the following expressions for the VBA:

$$\left\{\begin{array}{l} p(\boldsymbol{g}|\boldsymbol{f},\tau_\epsilon) = \mathcal{N}(\boldsymbol{g}|\boldsymbol{H}\boldsymbol{f},(1/\tau_\epsilon)\boldsymbol{I}) \\ p(\tau_\epsilon|\alpha_{\tau0},\beta_{\tau0}) = \mathcal{G}(\tau_\epsilon|\alpha_{\tau0},\beta_{\tau0}) \\ p(f|\tau) = \prod_j \mathcal{N}(f_j|0,1/\tau_j) \\ p(\tau|\alpha_0,\beta_0) = \prod_j \mathcal{G}(\tau_j|\alpha_0,\beta_0) \\ \\ q_1(f|\widetilde{\boldsymbol{\mu}},\widetilde{\boldsymbol{\Sigma}}) = \mathcal{N}(f|\widetilde{\boldsymbol{\mu}},\widetilde{\boldsymbol{\Sigma}}) \\ \widetilde{\boldsymbol{\mu}} = \langle\lambda\rangle\, \widetilde{\boldsymbol{\Sigma}}\boldsymbol{H}'\boldsymbol{g} \\ \widetilde{\boldsymbol{\Sigma}} = (\langle\lambda\rangle\, \boldsymbol{H}'\boldsymbol{H} + \widetilde{Z})^{-1}, \\ \text{with } \widetilde{Z} = \widetilde{T}^{-1} = \mathrm{diag}\,[\widetilde{\tau}] \end{array}\right. \quad \left\{\begin{array}{l} q_{2j}(\tau_j) = \mathcal{G}(\tau_j|\widetilde{\alpha}_j,\widetilde{\beta}_j) \\ \widetilde{\alpha}_j = \alpha_{00} + 1/2 \\ \widetilde{\beta}_j = \beta_{00} + < f_j^2 > /2 \\ \\ q_3(\tau_\epsilon) = \mathcal{G}(\tau_\epsilon|\widetilde{\alpha}_{\tau_\epsilon},\widetilde{\beta}_{\tau_\epsilon}), \\ \widetilde{\alpha}_{\tau_\epsilon} = \alpha_{\tau0} + (n+1)/2 \\ \widetilde{\beta}_{\tau_\epsilon} = \beta_{\tau0} + 1/2[\|\boldsymbol{g}\|^2 \\ -2\,\langle f\rangle'\,\boldsymbol{H}'\boldsymbol{g} + \boldsymbol{H}'\,\langle f f'\rangle\,\boldsymbol{H}] \end{array}\right. \quad \left\{\begin{array}{l} < f >= \widetilde{\boldsymbol{\mu}} \\ < f f' >= \widetilde{\boldsymbol{\Sigma}} + \widetilde{\boldsymbol{\mu}}\widetilde{\boldsymbol{\mu}}' \\ < f_j^2 >= [\widetilde{\boldsymbol{\Sigma}}]_{jj} + \widetilde{\mu}_j^2 \\ \widetilde{\lambda} = \widetilde{\alpha}_\tau / \widetilde{\beta}_\tau \\ \widetilde{\tau}_j = \widetilde{\alpha}_j / \widetilde{\beta}_j \end{array}\right. \tag{18}$$

Journal of Physics: Conference Series **386** (2012) 012006 doi:10.1088/1742-6596/386/1/012006

We can also express the free energy expression $\mathcal{F}(q)$ which can be used as a stopping criterion for the algorithm. Its expression is given in the appendix. The resulting algorithm can be summarized as follows

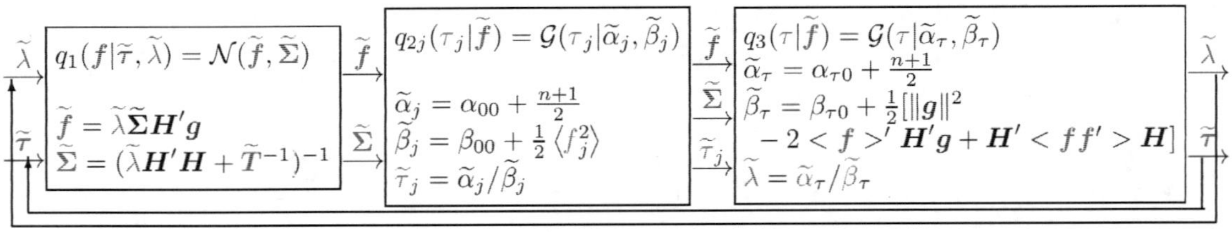

4. Bayesian Variational Approximation with Mixture of Gaussians priors

The mixture models are also very commonly used as prior models. In particular the Mixture of two Gaussians (MoG2) model:

$$p(\boldsymbol{f}|\lambda, v_1, v_0) = \prod_j (\lambda\mathcal{N}(f_j|0, v_1) + (1-\lambda)\mathcal{N}(f_j|0, v_0)) \tag{19}$$

which can also be expressed through the binary valued hidden variables $z_j \in \{0,1\}$

$$\left\{ \begin{array}{ll} p(\boldsymbol{f}|\boldsymbol{z}) = \prod_j p(f_j|z_j) & = \prod_j \mathcal{N}\left(f_j|0, v_{z_j}\right) \propto \exp\left\{-\frac{1}{2}\sum_j \frac{f_j^2}{v_{z_j}}\right\} \\ P(z_j = 1) = \lambda, & P(z_j = 0) = 1 - \lambda \end{array} \right. \tag{20}$$

In general $v_1 >> v_0$ and λ measures the sparsity $(0 < \lambda << 1)$ [10]. In this case also all the equations are very similarly can be obtained. Here, we do not have enough place to write them. They will be provided in the final version as an appendix.

5. Conclusions

In this paper, a VBA method is proposed for doing Bayesian computations for inverse problems where a hierarchical prior is used. In particular, two prior models are considered: the Student-t and the mixture of Gaussian models. In both cases, these priors can be written via hidden variables which gives the model a hierarchical structure which is used to do the factorization. All the details are given in the appendix. For some applications see for example [11, 12].

6. References

[1] Choudrey R A 2002 *Variational Methods for Bayesian Independent Component Analysis* Ph.D. thesis University of Oxford
[2] Beal M 2003 *Variational Algorithms for Approximate Bayesian Inference* Ph.D. thesis Gatsby Computational Neuroscience Unit, University College London
[3] Likas A C and Galatsanos N P 2004 *IEEE Transactions on Signal Processing*
[4] Winn J, Bishop C M and Jaakkola T 2005 *Journal of Machine Learning Research* **6** 661–694
[5] Chatzis S and Varvarigou T 2009 *IEEE Trans. on Fuzzy Systems* **17** 505–517
[6] Park T and Casella G 2008 *Journal of the American Statistical Association*
[7] Tipping M 2001 *Journal of Machine Learning Research*
[8] He L, Chen H and Carin L 2010 *IEEE Signal. Proc. Let.* **17** 233–236
[9] Fraysse A and Rodet T 2011 *SSP 2011* S17.5 (Nice, France) pp 605–608
[10] H Ishwaran J R 2005 *Annals of Statistics*
[11] Zhu S, Mohammad-Djafari A, Wang H, Deng B, Li X and Mao J 2012 *Eurasip Journal of Signal Processing* **special issue "sparse approximations in signal and image processing"**
[12] Chu N, Picheral J and Mohammad-Djafari A Bilbao, Spain, Dec14-17,2011 *IEEE International Symposium on Signal Processing and Information Technology* pp 286–289

Electromagnetic Inverse-Scattering Reconstructions by an Inexact Newton Method: Numerical and Experimental Results

Matteo Pastorino

Department of Naval, Electrical, Electronic and Telecommunication Engineering (DITEN) – University of Genoa

Via Opera Pia, 11A, I-16145, Italy

matteo.pastorino@unige.it

Abstract. This paper deals with the inverse scattering problem for electromagnetic imaging. In particular, a reconstruction procedure based on the inexact Newton method is reviewed. A new formulation, developed in the L^p functional Banach space, is also considered. Numerical and experimental results recently obtained by using the above mentioned reconstruction methods are discussed with specific reference to nondestructive evaluation of dielectric materials such as plastic and wood.

1. Introduction

In this paper, the application of the so-called inexact Newton method (INM) [1]-[6] to electromagnetic imaging is reviewed. The electromagnetic inverse scattering problem [7] represents the basic formulation for image-based methods devoted to the inspection of unknown bodies in several applications, ranging from nondestructive testing in industrial areas [8][9] to biomedical diagnostics [10]. When microwaves are used as interrogating waves, the field is scattered by the target and simplifying assumptions (e.g., ray propagation) cannot be applied. Consequently, the inverse scattering problem must be addressed starting from its "exact" equations [7][11]. Only in the case one knows some a priori information about the target some approximations can be applied (e.g., Born or Rytov approximations [12]). In general, however, the problem must be addressed by considering the nonlinear equations governing the scattering phenomenon.

In the following, the formulation of the considered inverse scattering problem will be briefly outlined in Section 2. We consider here the inspection of penetrable targets by using a two-dimensional tomographic configuration. In Section 3, the INM will be described, with particular emphasis to new developments in the L^p functional Banach space [13]-[17].

Numerical and experimental results obtained with the described approaches will be reviewed in Section 3. Finally, some concluding remarks are reported in Section 4.

2. Problem formulation

The inspection of a target as the solution of an inverse scattering problem can be performed following various approaches. First of all, the problem can be faced in time domain (for example, by using efficient time reversal techniques [18]) or in the frequency domain. In the last case, integral equations are usually solved [7].

We consider here the inspection of dielectric targets in a tomographic configuration. The scatterer is assumed to be an infinite cylinder illuminated under transverse-magnetic (TM) conditions. The cross section of the cylinder is arbitrarily shaped and inhomogeneous. A *exp*[$j\omega t$] time-dependence is assumed and omitted. The object cross section is represented by a contrast function

$$c(\mathbf{r}) = \varepsilon_r(\mathbf{r}) - 1 - \frac{j\sigma(\mathbf{r})}{\omega\varepsilon_0} \tag{1}$$

which describes the dielectric characteristics of the bounded investigation domain, It depends on the relative dielectric permittivity and on the electric conductivity [S/m] of the object ($\mathbf{r}$ denotes the position vector in the transversal plane and ω the angular frequency). The object is successively illuminated by incident waves and the field scattered is collected in a set of measurement points (the *observation domain,* D_{meas}). The inverse problem can be formulated by using the following two equations:

$$e_{scatt}(\mathbf{r}) = -k_0^2 \int_{D_{inv}} c(\mathbf{r}')e_{tot}(\mathbf{r}')g_0(\mathbf{r},\mathbf{r}')d\mathbf{r}', \quad \mathbf{r} \in D_{meas} \tag{2}$$

$$e_{tot}(\mathbf{r}) = e_{tot}(\mathbf{r}) + k_0^2 \int_{D_{inv}} c(\mathbf{r}')e_{tot}(\mathbf{r}')g_0(\mathbf{r},\mathbf{r}')d\mathbf{r}', \quad \mathbf{r} \in D_{inv} \tag{3}$$

where the involved field quantities refer to the incident, scattered, and total field components, which are polarized as the cylinder axis. The above equations are scalar ones, due to the assumptions about the problem geometry and illumination (the index of the multiple illumination is omitted here). Equation (2) is usually called the *data equation*, whereas equation (3) is called the *state equation* [7]. Both equations involve the Green's function for free space $g_0(\mathbf{r},\mathbf{r}')$ [12] and the propagation constant $k_0^2 = \omega^2\varepsilon_0\mu_0$. The integrals are performed over the *investigation domain* D_{inv}. Equations (2) and (3) can be solved simultaneously or they can be combined together in a unique operator equation having the contrast function as the sole unknown term, i.e.,

$$e_{scatt}(\mathbf{r}) = F(c) = G_{data}\tilde{c}((I - G_{state}\tilde{c})^{-1}e_{inc})(\mathbf{r}), \quad \mathbf{r} \in D_{meas} \tag{4}$$

where G_{state} and G_{data} denote the operators related to equations (1) and (2), respectively, and $\tilde{c}$ is given by $\tilde{c}(\cdot)(\mathbf{r}) = c(\mathbf{r})(\cdot)(\mathbf{r})$.

It should be mentioned that other formulations are possible, e.g., by using the equivalent current density

$$j_{eq}(\mathbf{r}) = c(\mathbf{r})e_{tot}(\mathbf{r}) \tag{5}$$

as an unknown. This approach is usually referred as the *contrast source* method [19]. In this case, it results that equation (2) is linear and some *mitigation* of the overall nonlinearity of the problem seems to be achievable [20]-[22].

Several numerical methods, both deterministic and stochastic ones have been devised to solve the above problem (see, for example, [23]-[41] and the references therein). In [42]-[48] the use of the INM has been proposed. The INM solves the nonlinear integral equations in a regularized way by "combined iterative processes with Newton's method as primary method and some, as yet unspecified, secondary method for solving the [obtained] linear systems" [3]. In particular, the scheme proposed in [42] is an outer/inner iterative algorithm which can be summarized as follows:

1) *Outer steps*: the equations are linearized by means of a Newton's first-order expansion involving the Fréchet derivative of the operator. The functional equation $F(x) = y$ (e.g., equation

(4)) is iteratively solved by finding a function h such that $F'_{xk} h = y - F(x_k), k = 0,1\dots$ (until a stopping criterion is fulfilled) and by updating the solution with $x_{k+1} = x_k + h$. F'_{xk} denotes the Fréchet derivative.

2) *Inner steps*: regularized solutions of the obtained linear equations are obtained by the truncated Landweber method [49]-[51]. In particular, this method is based on the following relation

$$h_{i+1} = h_i - \tau_k F'^{*}_{xk} (y_k - F'_{xk} (h_i)), i = 1,2,\dots, i_{\max} \tag{6}$$

where $0 < \tau_k < 2(\|F'^{*}_{xk} F'_{xk}\|_S^2)^{-1}$ is the relaxation parameter, $y_k = y - F(x_k)$, and $h_0 = 0$.

The inner iterative method is a regularization algorithm, i.e., the components of the input data corrupted by noise, usually related to the highest frequencies Fourier components, are filtered out during the first iterations [52]. A regularized solution must be computed since any Newton step gives rise to an ill-posed linear equation. A similar outer–inner Newton–Landweber scheme for a general nonlinear operator equation has first been introduced and comprehensively analyzed in [53]. In applying the INM, we explore the possibility of keeping fixed the number of inner iterations. In this way, the only parameter controlling the so-called semiconvergence [52] is the number of outer iterations.

The Landweber method is an iterative method endowed with very high regularization capabilities [51][52]. Although it is not fast as other methods, the Landweber method yields very high stability [52] and becomes competitive for solving nonlinear and severely ill-posed problems [2].

Although several interesting results have been obtained by applying the approach in [42] by using both simulated and experimental data (as described in Section 3), the considered version of the algorithm, as most of other regularization methods currently applied in electromagnetic imaging, minimizes the residual function (in Hilbert space) $R_2(x) = \|F(x) - y\|_{L^2}^2$. Recently, a different approach has been proposed [54]. In particular, a different cost functional is assumed, i.e., $R_p(x) = \|F(x) - y\|_{L^p}^r$ where $\|\cdot\|_{L^p}^r$ is the norm of the L^p functional Banach space. The metric of the Banach space L^p emphasizes, for values of the constant $1 < p < 2$, the points where the residual is small with respect to the classical L^2 norm. This choice gives rise to a substantial reduction of the over-smoothing in the restored solution x and has been proven to provide more accurate reconstructions. However, the INM required significant modifications, which are detailed in [54]. In particular, the inner loop is based on the following recursive relation ($h_0 = \tilde{h}_0^* = 0$)

$$\tilde{h}_{i+1}^* = \tilde{h}_i^* - \tau_k F_{xn}^* J_r^{L^p} (F_{xn} h_{n,k} - y_k), \tag{7}$$

$$h_{i+1} = J_{r*}^{L^{p*}} (\tilde{h}_{i+1}^*), i = 0,1,\dots, i_{\max} \tag{8}$$

where $J_r^{L^p} : L^p(D_{meas}) \to L^{p^*}(D_{meas})$ and $J_{r*}^{L^{p*}} : L^{p^*}(D_{meas}) \to L^p(D_{meas})$ are the duality maps of $L^{p^*}(D_{meas})$ and $L^p(D_{meas})$ [13] and $0 < \tau_k < 2(\| F'^{*}_{xk} F'_{xk} \|_{r*}^2)^{-1}$.

3. Review of numerical and experimental results.

The proposed approach has been intensively studied by using numerical methods. Results can be found in the mentioned references. Concerning in particular the improvement related to the new version developed in the L^p Banach space, it has been shown [54] that, for p = 1.2, we can expect a reduction of the mean reconstruction error on the contrast function inside the investigation domain of about 30% when compared with the same solution obtained in the Hilbert space (for a homogeneous cylinder with a triangular cross section of side 0.5λ, λ being the wavelength of the incident radiation, c = 1-j0.3, 8 line-current sources, 241 measurement points, and a signal to noise (SNR) ratio on the simulated scattered field of 20 dB). More relevant, the improvement becomes more and more evident when the noise on the data increases. In particular, the same error reduces of about 55% for SNR = 10 dB and about 60% for SNR = 5 dB. The new approach significantly affects the reconstruction of sharp scatterers, as the ones encountered, for example, in nondestructive testing applications. As shown in

[55], for a square cylinder with dielectric parameters similar to the ones of a cement paste, in which a circular crack is present, not only the overall quality is improved comparing with the standard approach, but the shape of the small defect is retrieved with a better accuracy, confirming that the metrics of the L^p Banach space, for p less than 2 (the reconstructed images reported in [56] concerns the case $p = 1.2$), allows for a certain *edge-preserving* effect, even if no regularizing terms of this kind (adopted by certain authors, e.g., in [56]) are used.

The INM has been also tested against experimental data. In [57], the measured data obtained at the Frésnel Institute of Marseille, France, have been successfully inverted by using the standard approach [42]. For the configuration composed by two adjacent circular cylinders (radii: $R_1 = 30$ mm and $R_2 = 80$ mm; $\varepsilon_{r1} = 3.0$ and $\varepsilon_{r2} = 1.45$; frequency $f = 2$ GHz), the new approach developed in the L^p Banach space has resulted in an improvement of more than 10% in the relative error on the reconstructed contrast function [54]. The original method has been also applied, in the framework of a joint research between DITEN, University of Genoa, Italy, and SUPSI, University of Lugano-Manno, Switzerland, to data measured by the microwave tomographic prototype described in [58]. In particular, wood and plastic (including hollowed cylinders filled with sand) objects have been inspected. Results can be also found in [59]-[61]. Quite surprisingly, good results have been recently obtained also in the case in which metallic inclusions are present in the dielectric structure (this fact has a significant importance in industry applications in which foreign bodies can affect the production process and even damage the used machinery) [62]. Although highly underestimated, the retrieved distribution of the electric conductivity allows one to precisely localize the inclusion, although no information about the high conductivity of the inspected target has been inserted into the model, Future work will be aimed at assessing the possible improvement that can be expected to these results when the new version of the INM is applied. In particular, it will be of interest to evaluate whether the same quality levels of the reported reconstructions can be obtained by a reduced number of views, with a corresponding reduction of the overall acquisition time.

4. Conclusions.

In this paper, the application of an inexact Newton method to the integral equations of the inverse scattering problem for electromagnetic imaging has been discussed. After a description of the considered tomographic configuration, the application of the inexact Newton method has been described, also with reference to a new formulation developed in the the L^p functional Banach space. Considering the numerical and experimental results obtained by applying this regularization scheme, the approach can be considered quite satisfactory and further investigations are expected in the light of practical applications in the field of nondestructive testing. In particular, the results obtained by using the new formulation, although preliminary, seems to indicate that a significant improvement in the reconstruction accuracy can be obtained, especially for measurements performed in a noisy environment.

5. Acknowledgment.

The author wish to thank Professors Claudio Estatico (Insubria University, Italy), Andrea Salvadè, Ricardo Monleone and co-workers (SUPSI, University of Lugano-Manno, Switzerland), and Andrea Randazzo (DITEN, University of Genoa, Italy). The reported material is in fact a part of a joint research with their scientific groups.

6. References.

[1] Dembo R S, Eisenstat S C and Steihaug T 1982 *SIAM J. Numer. Anal.* **19** 400–8.
[2] Rieder A 1999 *Inv. Probl.* **15** 309–27.
[3] Rheinboldt W C 1998 *Methods for Solving Systems of Nonlinear Equations* (Philadelphia:

SIAM Publications) Regional Conf. Series in Appl. Math. **70**.

[4] Eisenstat S C and Walker H F 1994 *SIAM J. Optim.* **4** 393–422.

[5] Eisenstat S C and Walker H F 1996 *SIAM J. Sci. Comput.* **17** 16–32.

[6] Liu K, Tian J, Yang X, Qin C, Zhu S, Han D and Ma X 2010 *Proc. Int. Conf. Engineering in Medicine and Biology Society (EMBC)* pp 3009-12.

[7] Pastorino M 2010 *Microwave Imaging* (Hoboken: John Wiley & Sons).

[8] Lesselier D and Bowler J 2002 Special issue on electromagnetic and ultrasonic nondestructive evaluation *Inv. Probl.* **18**.

[9] Chen C C, Johnson J T, Sato M and Yarovoy A G 2007 Special issue on subsurface sensing using ground-penetrating radar *IEEE Trans. Geosci. Remote Sens.* **45**.

[10] Zhang Z Q and Liu Q H 2004 *IEEE Trans. Biomed. Eng.* **51** 544–48.

[11] Colton D and Kress R 1998 *Inverse Acoustic and Electromagnetic Scattering Theory* (Berlin: Springer).

[12] Morse P M and Feshbach H 1953 *Methods of Theoretical Physics* (New York: McGraw-Hill).

[13] Schöpfer F, Louis A K and Schuster T 2006 *Inv. Probl.* **22** 311-29.

[14] Kaltenbacher B, Schöpfer F and Schuster T 2009 *Inv. Probl.* **25**.

[15] Daubechies I, Defrise M and De Mol C 2004 *Comm. on Pure and Applied Math.* **57** 1413-57.

[16] Canuto C and Urban K 2005 *SIAM J. Numer. Anal.* **42** 2043-75.

[17] Hein T and Kazimierski K S 2010 *Inv. Probl.* **26**.

[18] Gdoura S and Lesselier D 2012 *Proc. 2nd Int. Workshop on New Comput. Meth. for Inv.Probl. (Cahan).*

[19] van den Berg P M and Kleinman R E 1997 A contrast source inversion method *Inv. Probl.* **13**. 1607-20.

[20] Abubakar A, van den Berg P M and Mallorqui J J 2001 *IEEE Trans. Microwave Theory Tech.* **50** 1761-71.

[21] Gilmore C, Mojabi P and LoVetri J 2009 *IEEE Trans. Antennas Propagat.* **57** 2341-51.

[22] Abubakar A, van den Berg P M and Mallorqui J J 2001 *IEEE Trans. Microwave Theory Tech.* **50** 1761-71.

[23] Rubæk T, Meaney P M, Meincke P and Paulsen K D 2007 *IEEE Trans. Antennas Propagat.* **55** 2320-31.

[24] De Zaeytijd J, Franchois A, Eyraud C and Geffrin J-M *IEEE Trans. Antennas Propagat.* **55** 3279-92.

[25] Amineh R K, Ravan M, Trehan A and Nikolova N K 2011 *IEEE Trans. Antennas Propagat.* **59** 928–40.

[26] Giakos G C, Pastorino M, Russo F, Chowdhury S, Shah N and Davros W 1999 *IEEE Instrum. Meas. Mag.* **2** 32–35.

[27] Chew W C, Wang Y M, Otto G, Lesselier D and Bolomey J C 1994 *Inv. Probl.* **10** 547-52.

[28] Geffrin J M, Sabouroux P and Eyraud C 2005 *Inv. Probl.* **21** S117-30.

[29] Belkebir K, Kleinman R E and Pichot C 1997 *IEEE Trans. Microwave Theory Tech.* **45** 469–76.

[30] Micolau G, Saillard M and Borderies P 2003 *IEEE Trans. Geosci. Remote Sens.* **41** 1813–20,

[31] Ravat C and Joubert P Y 2009 *Sensor Lett.* **7** 400-5.

[32] Xu L, Davis S K, Hagness S C, van der Weide D W and Van Veen B D 2004 *IEEE Trans. Microwave Theory Tech.* **52** 1856–65.

[33] Tijhuis G, Belkebir K, Litman A and de Hon B 2001 *IEEE Trans. Geosci. Remote Sens.* **39** 1316–30.

[34] El-Shenawee M, Rappaport C, Miller E L and Silevitch M B 2001 *IEEE Trans. Geosci. Remote Sens.* **39** 1174–82.

[35] Lambert M and Lesselier D 2000 *Inv. Probl.* **16** 563–76.

[36] Benedetti M, Donelli M, Franceschini G, Pastorino M and Massa 2005 *IEEE Trans. Geosci. Remote Sens.* **43** 2584-92.

[37] Caorsi S, Massa A, Pastorino M, Raffetto M and Randazzo A 2003 *IEEE Trans. Antennas*

Propagat. **51** 2878-84.

[38] Caorsi S, Massa A and Pastorino M 2000 *IEEE Trans. Microwave Theory Tech.* **48** 1815-30.

[39] Brignone M, Bozza G, Randazzo A, Aramini R, Piana M and Pastorino M 2008 *IEEE Trans. Antennas Propagat.* **56** 3224-32.

[40] Franchois A, Joisel A, Pichot C and Bolomey J C 1998 *IEEE Trans. Med. Imag.* **17** 550–61.

[41] Kooij B J, Lambert M and Lesselier D 1999 *Radio Sci.* **34** 1361-71.

[42] Estatico C, Pastorino M and Randazzo A 2005 *IEEE Trans. Geosci Remote Sens.* **43** 2593-05.

[43] Estatico C, Bozza G, Pastorino M and Randazzo A 2006 *IEEE Antennas Wireless Propagat. Lett.* **5** 61-64.

[44] Bozza G, Estatico C, Pastorino M and Randazzo A 2007 *IEEE Geosci. Remote Sens. Lett.* **4** 51-55.

[45] Bozza G, Estatico C, Pastorino M and Randazzo A 2007 *Mat. Eval.* **65** 917-22.

[46] Di Benedetto F, Estatico C, Nagy J G and Pastorino M 2009 *Electronic Trans. Num. An. (ETNA)* **33** 105-125.

[47] Bozza G and Pastorino M 2009 *IEEE Trans. Antennas Propagat.* **57** 1122-32.

[48] Pastorino M 2011 *Proc. Ass. Gén. GDR Ondes: Interferences d'ondes (Nice).*

[49] Engl E, Hanke M and Neubauer A 2000 *Regularization of Inverse Problems* (Dordrecht: Kluwer Academic).

[50] Hanke M 1991 *Numer. Math.* **60** 341–373.

[51] Hanke H M, Neubauer A and Scherzer O 1999 *Numer. Math.* **72** 21–37.

[52] Bertero M and Boccacci P 1998 *Introduction to Inverse Problems inImaging* (Bristol: IOP).

[53] Kaltenbacher B 1997 *Inv.Probl.* **13** 729–53.

[54] Estatico C, Pastorino M and Randazzo A 2012 *IEEE Trans. Antennas Propagat.* (in press).

[55] Randazzo A and Estatico C 2012 *Nondestr. Testing Eval.* (in press).

[56] Lobel P, Pichot C, Blanc-Feraud L and Barlaud M 1997 *IEEE Antennas Propagat. Mag* **39** 12-14.

[57] Estatico C, Bozza G, Massa A, Pastorino M and Randazzo A 2005 *Inv. Probl.* **21** S81-94.

[58] Pastorino M, Salvadè A, Monleone R, Bozza G and Randazzo A 2009 *IEEE Trans. Instrum. Meas.* **58** 2072-79.

[59] Monleone R, Pastorino M, Fortuny-Guasch J, Salvadè A, Bartesaghi T, Bozza G, Maffongelli M, Massimini A and Randazzo A 2012 *IEEE Trans. Instrum. Meas.* **61** 140-48.

[60] Pastorino M, Salvadè A, Monleone R, Bartesaghi T, Bozza G and Randazzo A 2007 *Proc. IEEE Int. Workshop Imaging Systems Tech. (Krakow).*

[61] Salvadè A, Pastorino M, Monleone R, Bozza G, Bartesaghi T, Maffongelli M and Massimini A 2010 *Proc. URSI Comm. B Int. Symp. on Electromagnetic Theory (Berlin).*

[62] Maffongelli M, Monleone R, Pastorino M, Poretti S, Randazzo A and Salvadè A 2012 *IEEE Microwave Wireless Comp. Lett.* (in press).

2nd Int. Workshop on New Computational Methods for Inverse Problems (NCMIP 2012) IOP Publishing
Journal of Physics: Conference Series **386** (2012) 012008 doi:10.1088/1742-6596/386/1/012008

Rao-Blackwellised Interacting Markov Chain Monte Carlo for Electromagnetic Scattering Inversion

F. Giraud[1], P. Minvielle[1], M. Sancandi[1], P. Del Moral[2,]

[1] CEA-CESTA, 33114 Le Barp, France

[2] INRIA Bordeaux Sud-Ouest, Domaine Universitaire, 351, cours de la Liberation, 33405 Talence Cedex, France

E-mail: francois.giraud@cea.fr

Abstract. The following electromagnetism (EM) inverse problem is addressed. It consists in estimating local radioelectric properties of materials recovering an object from the global EM scattering measurement, at various incidences and wave frequencies. This large scale ill-posed inverse problem is explored by an intensive exploitation of an efficient 2D Maxwell solver, distributed on High Performance Computing (HPC) machines. Applied to a large training data set, a statistical analysis reduces the problem to a simpler probabilistic metamodel, on which Bayesian inference can be performed. Considering the radioelectric properties as a dynamic stochastic process, evolving in function of the frequency, it is shown how advanced Markov Chain Monte Carlo methods, called Sequential Monte Carlo (SMC) or interacting particles, can provide estimations of the EM properties of each material, and their associated uncertainties.

1. Introduction

The inverse problem is described in figure 1. The Radar Cross Section quantifies the scattering power of an object, at a given incidence and wave frequency. It is defined as the ratio between the radar transmitted power and the incident power density (in plane wave) [1]. RCS measurement process is schematically presented on the right part of figure 1. The object or mock-up is illuminated by a quasi-planar monochromatic wave, inside an anechoic chamber where interferences are limited.The acquisitions are realized at K successive discrete frequencies $(f_1, f_2, \cdots, f_K)$, for different incidence angles (by piloting motorized rotating support). From this raw data, a signal processing is performed, mainly consisting of calibration and filtering. At the end, the measurement provides an evaluation of the calibrated complex (amplitude and phase) scattering coefficient, for each frequency and incidence.

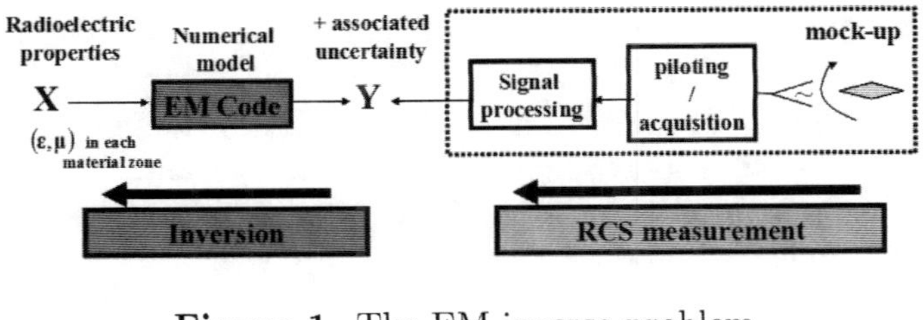

Figure 1. The EM inverse problem

A metallic axi-symmetric object is recovered with N areas (see figure 3), each area corresponding to a material with its associated isotropic radioelectric properties, i.e. the complex parameters of permittivity ϵ and permeability μ. The EM inverse problem can be expressed as: is it possible to extract some local information on the material properties (ϵ_k, μ_k) of each area from the global scattering measurement?

2. The stochastic modeling

At a given frequency f_k, the system state $\mathbf{X}_k$ can be defined by (omitting f_k to lighten the notations): $\mathbf{X}_k = \left[\underline{\epsilon}' \ \underline{\epsilon}'' \ \underline{\mu}' \ \underline{\mu}''\right]^T$, with $\underline{\epsilon}'$, $\underline{\epsilon}''$, $\underline{\mu}'$ and $\underline{\mu}''$, respectively the real and imaginary permittivity and permeability components of the N areas. Notice that the state dimension can be high ($4 \cdot N$ with $N \approx 100$). On the other hand, considering measurements at the given frequency f_k with various incidence angles in both polarizations (TM and TE), the observation vector is made of the real ($\Re(\cdot)$) and imaginary ($\Im(\cdot)$) parts of the complex scattering coefficients $\mathbf{c_{TM}}$ and $\mathbf{c_{TE}}$ measured at M angles $\theta_1, \cdots, \theta_M$: $\mathbf{Y}_k = [\Re(\mathbf{c_{TM}}) \ \ \Im(\mathbf{c_{TE}}) \ \ \Re(\mathbf{c_{TM}}) \ \ \Im(\mathbf{c_{TE}})]^T$. Considering all frequencies, vectors $\mathbf{X} = (\mathbf{X}_1, \ldots, \mathbf{X}_K)$ and $\mathbf{Y} = (\mathbf{Y}_1, \ldots, \mathbf{Y}_K)$ respectively define complete system state and observation.

2.1. Observation model

The 2D-axisymetric Maxwell solver software ($\mathcal{F}_{2D}$) can predict the observation from the system state. Assuming a multidimensional Gaussian measurement uncertainty model, it leads to the following likelihood model $\mathbf{Y}_k \mid \mathbf{X}_k \sim \mathcal{N}(\mathcal{F}_{2D}(\mathbf{X}_k), \mathbf{R}_k)$ (at a given frequency f_k), where $\mathbf{R}_k$ is the covariance matrix (assumed known). To avoid numerous and heavy $\mathcal{F}_{2D}$ computations, we have developed the following global approach. First, the high dimension state space and the associated system response are explored randomly around expected properties (prior knowledge), computations being massively distributed on HPC machines.

Let $\mathcal{B}_A = \{(\mathbf{X}_k^{(1)}, \mathbf{Y}_k^{(1)}), \cdots, (\mathbf{X}_k^{(N_E)}, \mathbf{Y}_k^{(N_E)})\}$ the training data composed of N_E couples. It is then processed by N-D statistical techniques; sensibility analysis and model reduction techniques can possibly reduce the state space dimension. Applying multidimensional regression, it turns out that the model is approximatively linear. According to the studied cases, the linearity errors, evaluated by residue analysis and bootstrap techniques, are significant but much lesser than the RCS measurement uncertainties. Finally, the following linear Gaussian model can be considered (at a given frequency f_k):

$$\mathbf{Y}_k \mid \mathbf{X}_k \sim \mathcal{N}(\mathbf{A}_k \cdot \mathbf{X}_k + \mathbf{Y}_k^0, \mathbf{R}_k) \tag{1}$$

where the deterministic part of the linear model is given by the learned matrix $4M \times 4N$ $\mathbf{A}_k$ and the vector $\mathbf{Y}_k^0$.

2.2. Prior model at a fixed frequency

Let us fix a frequency f_k. We model our a priori knowledge on $\mathbf{X}_k$ with a Gaussian distribution $\mathcal{N}(\mathbf{m}_k, \mathbf{P}_k)$. The object is divided in N_b blocks of areas, each of them composed of a rather homogeneous material. The location of these blocks is known exactly. Prior mean value $\mathbf{m}_k$ is defined with reference values of ϵ', ϵ'', μ' and μ'' for each of theses blocks. Then, for any component $\mathbf{X}_k^i$, we define a variance ${\sigma_{S,k}^i}^2$ as a mix between absolute and relative uncertainty. To take into account the spatial local homogeneity, covariance matrix $\mathbf{P}_k$ is defined block by block independently (and separately for each of the ϵ', ϵ'', μ', μ'') by correlation relations between $\mathbf{X}_k$'s block-sharing components:

$$Cov(\mathbf{X}_k^i, \mathbf{X}_k^j) = {\sigma_{k,S}^i}^2 \cdot \rho_S^{|i-j|}$$

where $\rho_S \in [0,1]$ is a spatial correlation parameter (typically $\rho_S = 0.95$ in our applications). It means that the correlation between 2 areas dicreases geometrically at speed ρ_S with the distance between them.

2.3. Inter-frequential prior model

Radioelectric properties ϵ', ϵ'', μ', μ'' are known to vary in function of the wave frequency [1]; their non-stationary dynamic can be quite different from frequency f_1 to f_K. However, in order to take account of expected frequency profiles regularity (for each material and ϵ', ϵ'', μ', μ''), we model sequence $(\mathbf{X}_k, k \in \{1,\ldots,K\}$ with a linear Gaussian correlation structure, given by the generalised autoregressive (AR) random process [2]:

$$\begin{aligned} &\mathbf{X}_1 \sim \mathcal{N}(\mathbf{m}_1, \mathbf{P}_1) \\ &(\mathbf{X}_{k+1} - \mathbf{m}_{k+1}) = \mathbf{D}_\rho \cdot \mathbf{H}_{k+1} \cdot \mathbf{H}_k^{-1} \cdot (\mathbf{X}_k - \mathbf{m}_k) + \sqrt{\mathbf{I_d} - \mathbf{D}_\rho^2} \cdot \mathbf{H}_{k+1} \cdot \mathbf{V}_k \end{aligned} \tag{2}$$

where

- for each k, $\mathbf{H}_k$ is the "square root" matrix of covariance matrix $\mathbf{P}_k$, defined as being the unique symetric definite positive matrix satisfying $\mathbf{H}_k \cdot \mathbf{H}_k^T = \mathbf{P}_k$;
- $(\mathbf{V}_k, k \in \{1,\ldots,K\})$ are i.i.d. $\mathcal{N}(0, \mathbf{I_d})$;
- $\mathbf{D}_\rho$ is a positive diagonal matrix commuting with the $\mathbf{H}_k$, further described, and depending on a parameter ρ.

One notices that so defined sequence $\mathbf{X}_k$ still admits $\mathcal{N}(\mathbf{m}_k, \mathbf{P}_k)$ as marginal distributions. More generally, it can be shown that the distribution of concatenated vector $\mathbf{X} = (\mathbf{X}_1, \ldots, \mathbf{X}_K)$ is a Gaussian distribution with mean $\mathbf{m} = (\mathbf{m}_1, \ldots, \mathbf{m}_K)$ and covariance matrix:

$$\mathbf{P} = \mathcal{H} \cdot \begin{pmatrix} \mathbf{I_d} & \mathbf{D}_\rho & \mathbf{D}_\rho^2 & \cdots & \mathbf{D}_\rho^{K-1} \\ \mathbf{D}_\rho & \mathbf{I_d} & \mathbf{D}_\rho & & \vdots \\ \mathbf{D}_\rho^2 & \mathbf{D}_\rho & \ddots & \ddots & \vdots \\ \vdots & & \ddots & \ddots & \mathbf{D}_\rho \\ \mathbf{D}_\rho^{K-1} & \cdots & \cdots & \mathbf{D}_\rho & \mathbf{I_d} \end{pmatrix} \cdot \mathcal{H}^T$$

where $\mathcal{H}$ is the diagonal by block matrix $\mathcal{H} = \mathrm{diag}(\mathbf{H}_1, \ldots, \mathbf{H}_K)$, which explicits every joint distribution $(\mathbf{X}_i, \mathbf{X}_j)$.

Finally, let us clarify parameter ρ (and matrix $\mathbf{D}_\rho$). $\mathbf{D}_\rho$'s role in these equations makes clear that it's a frequential correlation parameter. Its dimension is to be chosen among 3 possibilities according to our assumptions.

- 1st case: ρ is assumed independant from material and ϵ', ϵ'', μ', μ''. It is one-dimensional, $\rho \in [0,1]$, and $\mathbf{D}_\rho = \rho.\mathbf{I_d}$.
- 2nd case: ρ depends on the material (block). It is N_b-dimensional, $\rho \in [0,1]^{N_b}$, and $\mathbf{D}_\rho$ giving to each line (component of $\mathbf{X}_k$) its associated component of ρ.
- 3rd case: ρ depends on material and ϵ', ϵ'', μ', μ''. It is $4.N_b$-dimensional, and $\mathbf{D}_\rho$ is the diagonal matrix composed with $4.N_b$ $\rho_i.\mathbf{I_d}$-type blocks.

2.4. Global model

Conditionnaly to frequential correlation parameter ρ, the problem of the determination of $\mathbf{X} = (\mathbf{X}_1, \cdots, \mathbf{X}_K)$ given the measurements $\mathbf{Y} = (\mathbf{Y}_1, \cdots, \mathbf{Y}_K)$ can be expressed as a classic linear Gaussian hidden dynamic Markov process observed at "times" f_k $(k = 1, \cdots, K)$:

$$\mathbf{X}_{k+1} = \mathbf{M}_k^{\rho} \cdot \mathbf{X}_k + \mathbf{w}_k^{(1)} \quad \text{and} \quad \mathbf{Y}_k = \mathbf{A}_k \cdot \mathbf{X}_k + \mathbf{w}_k^{(2)}$$

where $\mathbf{w}_k^{(1)}$ and $\mathbf{w}_k^{(2)}$ are independant Gaussian noises with known parameters, and $\mathbf{M}_k^{\rho}$ a known matrix depending on ρ (see (1) and (2)). Parameter ρ, intuitively representing the inter-frequency regularity, is unknown and to be estimated. In respect to the probabilistic point of vue, it is probabilized, and given a prior distribution $p(\rho)$.

3. Sequential Monte Carlo approach for global inversion

3.1. Rao-Blackwellised SMC algorithm

The posterior distribution $p(\mathbf{X}, \rho|\mathbf{Y})$ can be decomposed as: $p(\mathbf{X}, \rho|\mathbf{Y}) = p(\mathbf{X}|\rho, \mathbf{Y}) \cdot p(\rho|\mathbf{Y})$. Conditionally to ρ, the system is linear Gaussian: the conditional distributions $p(\mathbf{X}|\rho, \mathbf{Y})$ can be straightforwardly computed by Kalman filtering, including in this off-line context backward Kalman smoothing. On the other hand, the term $p(\rho|\mathbf{Y}) \propto p(\mathbf{Y}|\rho) \cdot p(\rho)$ can be evaluated (up to a normalising constant) for a given ρ using the likelihood term provided by the Kalman filter and prior distribution $p(\rho)$. Consequently, in order to exploit this conditional structure of the system, Kalman smoothers are applied and integrated in an interacting particle approach. This idea of mixing analytic integration (here Kalman evaluation of $p(\mathbf{X}|\rho, \mathbf{Y})$) with stochastic sampling is a variance reduction approach, known as Rao-Blackwellisation [3].

Similarly to [3], we choose to implement an efficient interacting particle approach, in order to estimate the marginal distribution $\eta(\rho) := p(\rho|\mathbf{Y}) \propto p(\mathbf{Y}|\rho) \cdot p(\rho)$. Sequential Monte Carlo (SMC) is a stochastic algorithm to sample from complex high-dimensional probability distributions. The principle (e.g., [4]) is to approximate a sequence of target probability distributions (η_n) by a large cloud of random samples termed particles $(\zeta_n^k)_{1 \leq k \leq N_p} \in E^{N_p}$, E being called the state space. Between "times" $n-1$ and n, the particles evolve in state space E according to 2 steps:

- a **selection** step: every particle ζ_{n-1}^i is given a weight ω_i defined by a selection function g_n $(\omega_i = g_n(\zeta_{n-1}^i))$. By resampling (stochastic or deterministic), low-weighted particles vanish and are replaced by replicas of high-weighted ones.
- a **mutation** step: each selected particle $\widehat{\zeta}_{n-1}^i$ move, independently from the others, according to a Markov kernel M_n.

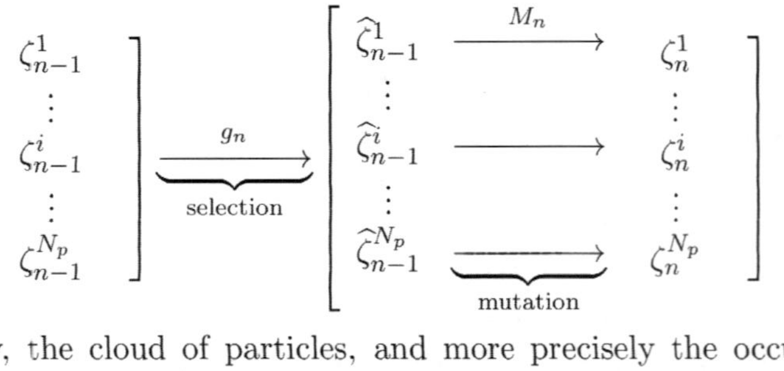

Evolving this way, the cloud of particles, and more precisely the occupation distribution $\eta_n^{N_p} = \frac{1}{N_p} \sum_{k=1}^{N_p} \delta_{\zeta_k^n}$ (sum of Dirac distributions), approximates for each n the theoretical distribution η_n defined recursively by the Feynman-Kac formulae, associated with the potentials g_n and kernels M_n [5].

Back to our objective of sampling from η, we then define the sequence of distributions $\eta_n(\rho)$:

$$\eta_n(\rho) \propto p(\mathbf{Y}|\rho)^{\alpha_n} \cdot p(\rho)$$

where $(\alpha_n)_{0 \leq n \leq n_f}$ is a sequence of number increasing from 0 to 1, so that: η_0 is prior distribution $p(\rho)$, easy to sample, η_{n_f} is target distribution η and sequence (η_n) admits a Feynman-Kac type structure with calculable selection functions g_n and Markov kernels chosen so that $\eta_n.M_n = M_n$ (Metropolis-Hastings for example). The distribution η is then interpreted as being the last distribution of a Feynman-Kac sequence, on which SMC can be performed, the estimator of η being the occupation distribution η^{N_p} extracted from the last cloud of particles.

3.2. Results

The occupation distribution η^{N_p} which approximates $\eta = p(\rho|\mathbf{Y})$ can be represented dimension by dimension via histograms (see figure 2).

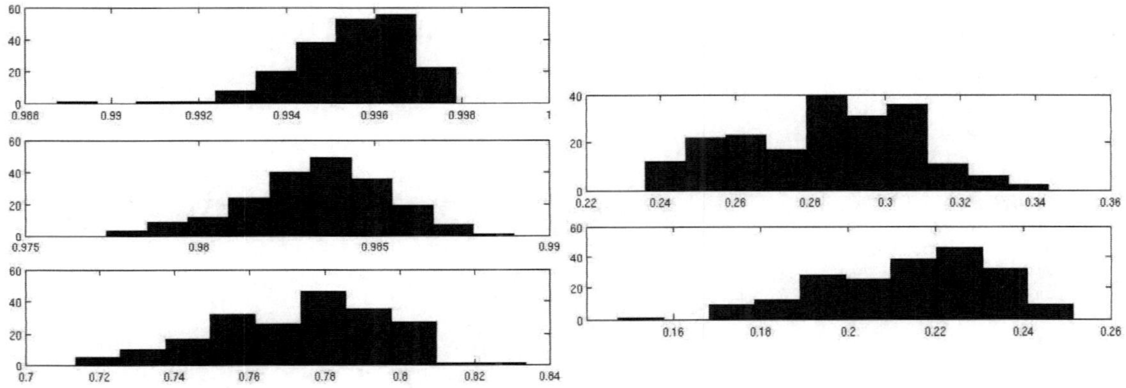

Figure 2. Estimation of a 5-dimensional ρ's distribution with $N_p = 100$ particles.

For each frequency f_k, this approximation $\eta^{N_p} \simeq \eta$, associated with the theoretical conditioning relations

$$E(\mathbf{X}_k|\mathbf{Y}) = E\left[E(\mathbf{X}_k|\rho, \mathbf{Y})|\mathbf{Y}\right]$$
$$Cov(\mathbf{X}_k|\mathbf{Y}) = E\left[Cov(\mathbf{X}_k|\rho, \mathbf{Y})|\mathbf{Y}\right] + Cov\left[E(\mathbf{X}_k|\rho, \mathbf{Y})|\mathbf{Y}\right]$$

can deliver estimators of respectively the mean and the covariance matrix of $\mathbf{X}_k$ .Roughly speaking, the posterior estimation is performed by randomly picking a ρ_i from the final cloud of particles and computing associated samples of $\mathbf{X}_k$ by a Kalman smoother conditionally to ρ_i. It is illustrated in figure 3, with a good agreement between the true state and estimated state.

Moreover, for any fixed area, the method provides estimators of the mean and marginal variance for every frequency, so that the results can be presented as frequential profiles, with marginal uncertainties (see figure 4). Even when the true (simulated) EM property profiles are chosen markedly divergent from the prior AR-type model, the method turns out to be robust. It results from the adaptive estimation of ρ which provides small values of ρ (i.e. weak correlation of EM properties for close frequencies) in the case of highly irregular true profiles.

4. Conclusion

An efficient statistical inference approach has been applied to estimate local material radioelectric properties from global EM scattering measurements. It combines intensive computations, meta-modeling and advanced sequential Monte Carlo techniques dedicated to frequency dynamic estimation.

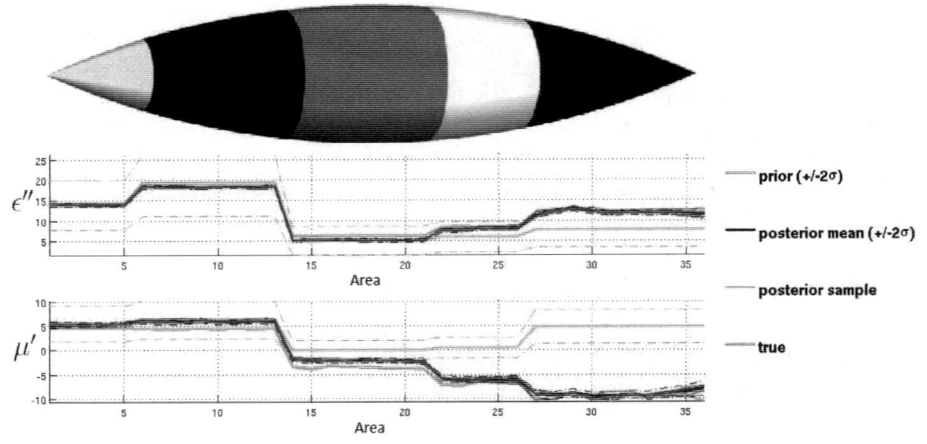

Figure 3. Estimated radioelectric properties of a $N_b = 5$-block object ($N = 19$ areas), at fixed frequency f_k.

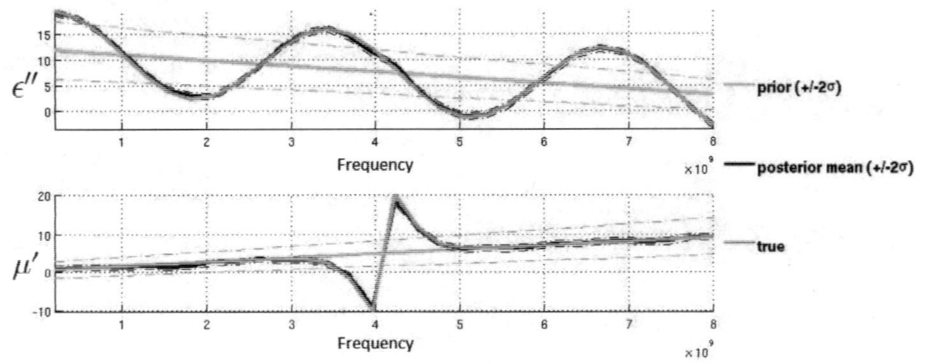

Figure 4. Estimated radioelectric properties ($K = 30$ frequencies) of a fixed area.

References

[1] E. F. Knott, *Radar Cross Section Measurements*, SciTech Publishing, 2006.

[2] S. M. Pandit, S-M Wu, *Time Series and System Analysis with Applications*, John Wiley & Sons, 1983.

[3] J. S. Liu and R. Chen, *Sequential Monte Carlo for Dynamic Systems*, Journal of the American Statistical Association, 93, 1032-1044, 1998.

[4] P. Del Moral, A. Doucet and A. Jasra, *Sequential Monte Carlo Methods for Bayesian Computation*, Bayesian Statistics 8, Oxford University Press, 2006.

[5] P. Del Moral, *Feynman-Kac Formulae, Genealogical and Interacting Particle Approximations*, Springer New York (Series: Probability and Applications), 2004.

Journal of Physics: Conference Series **386** (2012) 012009 doi:10.1088/1742-6596/386/1/012009

A reduced physical optics model for electromagnetic scattering problem

M. Amirisazi, F. Daout and F. Schmitt

SATIE, ENS Cachan, CNRS, UniverSud, 61 av du Président Wilson F-94235 Cachan cedex, France

E-mail: amiri@satie.ens-cachan.fr, fdaout@u-paris10.fr, fschmitt@u-paris10.fr

Abstract. The ultimate simplified *point scatterer model* contains no object's geometry informations, so a more complex prediction model containing geometrical informations is needed. The *Physical Optics* (PO) approximation has been widely used and considered as a good approximation of the far field electromagnetic scattering. Here, a forward model of bistatic scattering by PEC infinite elliptic cylinder based on the Physical Optics approximation is presented. The aim is to develope a simple scattering model for 2D targets illuminated by a monochromatic wave bistatic Radar. The reduced model is obtained by applying the stationary phase method to the PO integrals in both TE and TM wave cases. A parameter estimation procedure is also presented in order to examinate the efficiency of the presented model.

1. Introduction

The objective of the presented study is to develope a simple scattering model for 2D PEC objects to extract geometrical informations (as shape and size) using Radar's data, which is not available by some classical models as the point scatterer model [1].

The extraction of the approximated forward model is briefly presented in Section 2. The stationary phase method is applied to the PO integral equations which yields a approximated physical optics far-field direct scattering model. In Section 3, the validity of the presented model is confirmed by the comparison between the approximated model and results based on the Method of Moments (MoM) [2]. In Section 4, the parametric model is exploited in the parameter estimation methods, Adaptive Matched Filter (AMF) [3] and Minimum Mean Square Error (MMSE) [4] estimator, in order to extract the object's geometrical characterizations. Section 5, presents an application of presented method which allows to characterize the geometrical parameters of the objects.

2. Forward model extraction

Considering the geometry shown in Fig. 1, where the surface S of scatterer is shift invariant along the z-axis, the surface integral of the EFIE and the MFIE, can be reduced to [2]

$$A = A^{inc} + \oint_C \left[A\frac{\partial G}{\partial \widehat{n}} - G\frac{\partial A}{\partial \widehat{n}} \right] \mathrm{d}l \qquad (1)$$

where C is the cross-sectional contour of S and $\widehat{n}$ is the surface unit outward normal.

The parameter A is replaced by $E(\rho, \varphi)$ or $H(\rho, \varphi)$ in cylindrical coordinates and the time

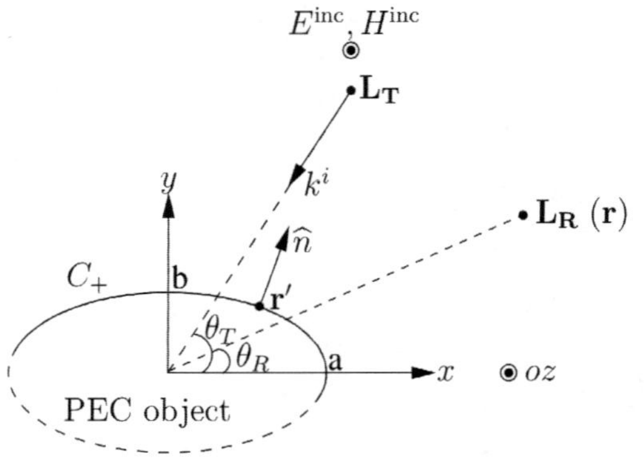

Figure 1. Scattering geometry by an infinite cylinder

dependence of the field is assumed to be $\exp(-j\omega t)$, where ω is the angular frequency. The function G is the well known two dimensional frequency dependent Green's function in free space, which satisfies the Helmholtz equation and has the form[5]

$$G(\mathbf{r}, \mathbf{r}') = -\frac{j}{4} H_0^2(k\|\mathbf{r} - \mathbf{r}'\|) \ , \ \ \mathbf{r} \neq \mathbf{r}' \tag{2}$$

where k is the free space wave number and H_0^2 is the Hankel function of the second kind of zero order.

The electric and magnetic fields of the incident plane wave are considered as parallel to the axis of the cylinder, at the observation cylindrical coordinates $\mathbf{r}'$ they are expressed as follows

$$\mathbf{P}_z^i(\mathbf{L}_T, \mathbf{r}') = -\frac{j}{4} H_0^2(k\|\mathbf{L}_T - \mathbf{r}'\|)\, \widehat{z} \tag{3}$$

where $\mathbf{P}^i = \mathbf{E}_z$ for TM wave, $\mathbf{P}^i = \mathbf{H}_z$ for TE wave and $\mathbf{L}_T$ is the vector representing the position of the transmitter.
For significant values of $\{k\|\mathbf{L}_T - \mathbf{r}\|,\ k\|\mathbf{r} - \mathbf{r}'\|\} \gg 0$, Hankel function in (2) and (3) can be replaced by its large argument asymptotic form [6]

$$H_0^2(x) \sim \sqrt{\frac{2}{\pi x}} \exp(-jx + \frac{\pi}{4}) \qquad \text{as} \ \ |x| \to \infty \tag{4}$$

The Kirchhoff approximation [7], also known as the Physical-Optics, assumes that the surface S is replaced at each point by its tangent plane. This approximation is valid for smooth objects with radius of curvature κ larger than the investigating wavelength λ.

Assuming $\{\|\mathbf{L}_R\|,\ \|\mathbf{L}_T\|\} \gg \|\mathbf{r}'\|$ ($\mathbf{L}_R$ represents the position of the receptor and $\mathbf{r}'$ is on the geometrically illuminated side of the scatterer denoted by C_+), *i.e* far-field measurement and $\lambda \ll \kappa$, high-frequency measurement, yields

$$E^s_{\rm TM}(\mathbf{L}_R) = E_0\, \rho_0 e^{j\phi_0} \int_{C_+} \widehat{n}.\widehat{L}_T e^{jk(\widehat{L}_T + \widehat{L}_R).\mathbf{r}'} \Gamma(\mathbf{r}') d\mathbf{r}' \tag{5}$$

$$H^s_{\rm TE}(\mathbf{L}_R) = H_0\, \rho_0 e^{j\phi_0} \int_{C_+} \widehat{n}.\widehat{L}_R e^{jk(\widehat{L}_T + \widehat{L}_R).\mathbf{r}'} \Gamma(\mathbf{r}') d\mathbf{r}' \tag{6}$$

where $\int_{C_+}$ denotes an integral over the illuminated portion of the target with infinitesimal part of the curve $\Gamma(\mathbf{r}')d\mathbf{r}'$ and $\rho_0 e^{j\phi_0} = \frac{e^{-jk(\|\mathbf{L}_T\| + \|\mathbf{L}_R)\|)}}{8\pi\sqrt{\|\mathbf{L}_T\|.\|\mathbf{L}_R)\|}}$.

The stationary phase method [8] states that these oscillatory integrals have an evaluation containing the significant non-zero contributions that occur in regions of the integration range where rapidly-varying phase function vanishes, *i.e.* at points of stationary phase.
It states that an integral of the form:

$$I = \int_{-\infty}^{+\infty} \rho(r) e^{jk\Phi(r)}\, dr \tag{7}$$

has an approximation containing contributions from the critical points inside the boundary of integration

$$I \sim \left(\frac{2\pi}{k}\right)^{\frac{1}{2}} \sum_{i=1}^{n} \frac{e^{sign(\Phi''(r_i))\frac{\pi}{4}}}{\|\Phi''(r_i)\|^{\frac{1}{2}}} \rho(r_i)\; e^{jk\Phi(r_i)} \tag{8}$$

where r_i is a point of stationary phase witch satisfy the condition $\Phi'(r) = 0$.

The scatterer is an infinite PEC elliptic cylinder of radius a and b, and we suppose $E_0 = H_0 = 1$. Then using the stationary phase evaluation of scattered field for both TM and TE waves, yields

$$A_{TM,TE} = \rho_0 \rho_p e^{j(\phi_0 + \phi_p - \pi/4)} \tag{9}$$

where ρ_p and ϕ_p are function of $(\theta_T,\, \theta_R,\, a,\, b)$,

$$\rho_p = \frac{ab(1 + \cos\theta_T \cos\theta_R + \sin\theta_T \sin\theta_R)}{\left(a^2(\cos\theta_T + \cos\theta_R)^2 + b^2(\sin\theta_T + \sin\theta_R)^2\right)^{3/4}} \tag{10}$$

and

$$\phi_p = k\left(a^2(\cos\theta_T + \cos\theta_R)^2 + b^2(\sin\theta_T + \sin\theta_R)^2\right)^{1/2} \tag{11}$$

θ_R and θ_T are the reception and transmission azimuth.

In the general case, where the semi-axis of cylinder have the angle γ with x, y axis, (9) can be used by the transformation $\theta_T \rightarrow (\theta_T - \gamma)$ and $\theta_R \rightarrow (\theta_R - \gamma)$.

3. Forward Model validity

The numerical solution provided by MoM for TM polarization, is used as reference to verify the model, where 800 of segments on the object's surface ensures the segment length $\triangle\, l < \lambda/9$ of the object's circumference.
Fig. 2 shows the comparison between the MoM and stationary phase model for an elliptic cylinder of dimensions $a = 5\,\lambda$, $b = 20\,\lambda$, and simulation conditions given in Table 1.
In this case, the validity range is limited (due to PO approximation) to 112°, *i.e.* $|\theta_R - \theta_T| <$ 112°, for *Phase-error*$_{max} = \pi/4$ and *Amplitude-error*$_{max}$ = 10%. It shows that the model is a good approximation of scattered field outside of the forward-scattering region, $|\theta_R - \theta_T| \rightarrow 180°$.
Fig. 2 indicates the validity of the present approximation for the scattering by an elliptic cylinder in the mentioned hypothetical contexts of measurement. The phase prediction is more accurate than the amplitude due to stationary phase approximation. In spite of the reduced range of validity, the model behavior remains well approximated.

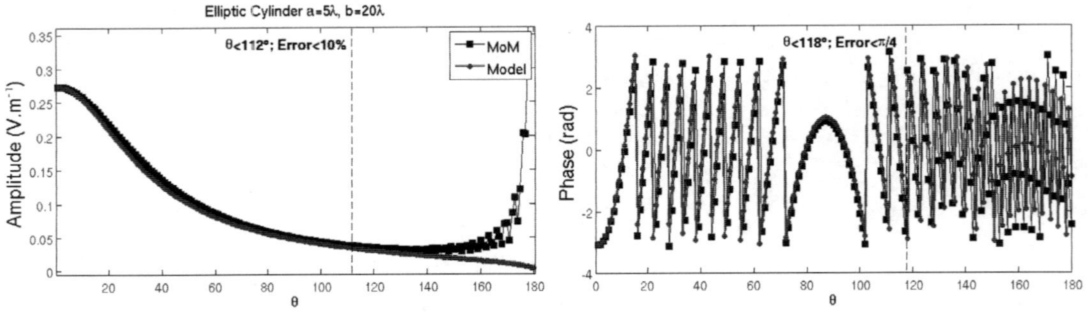

Figure 2. The reduced model compared with MoM; TM polarization, Elliptic cylinder $a = 5\lambda$, $b = 20\lambda$; $\lambda = 2$ cm;

4. Parameter estimation methods

By adaptive matching, the measured signal is correlated with the parameterized filter, here the target position is considered as known:

$$I_{\mathrm{AMF}}(a_i, b_j) = \| \sum_{\theta_R} \mathrm{S}(\theta_R).A^*(a_i, b_j, \theta_R) \| , \quad [\widehat{a}, \widehat{b}] = \{a, b \mid I_{\mathrm{AMF}}(a_i, b_j) \text{ is max}\} \tag{12}$$

where, S is the measured signal and A^* is the complex conjugate of A (eq. 9)
The product SA^*, is the effect of parameterized filter on the signal and is to be maximized in order to maximize the likelihood ratio. The approach consists of numerically searching the maximum value of two dimentional function I_{AMF} in the space of (a, b), where a and b are positive variables.

The Minimum mean square error estimator searches the parameters minimizing the Mean Square Error (MSE) between the measured signal and the response predicted by the model

$$I_{\mathrm{MMSE}}(a_i, b_j) = \sum_{\theta_R} \|\mathrm{S}(\theta_R) - A(a_i, b_j, \theta_R)\|^2 , \quad [\widehat{a}, \widehat{b}] = \{a, b \mid I_{\mathrm{MMSE}}(a_i, b_j) \text{ is min}\} \tag{13}$$

Values of MSE are used to comparative purpose and the model with lower MSE is considered as the best predictor model.

5. Application

The numerical results are obtained for the TM synthetic signal $\mathrm{S}(\theta_R)$, generated by Method of Moments then corrupted by a circular complex additive white Gaussian noise (SNR = 5 dB). Here, the electromagnetic interactions between the cylinders are neglected. The simulation conditions and scene configuration are listed in the Tables 1 and 2.

Table 3 represent the estimated radius of the simulation scene for MMSE and AMF methods. A comparison between the estimated parameters, indicate no remarkable difference between AMF and MMSE in this application and there is no performance degradation as a function of SNR, until the false alarms appears about SNR = 0 dB.

6. Conclusion

We have presented an approximated analytical scattering model of a PEC infinite elliptical cylinder. The model is valid under far-field and high-frequency measurement and it is used to generate signal model in the AMF and MMSE estimation methods.

Frequency(GHz)	f_0	14.3
Transmitter position	L_T	500 λ
	θ_T	0°
Receptor positions	L_R	500 λ
	θ_R	0°.. 360°
	$\Delta\, \theta_R$	1°

Table 1. Simulation conditions

Object	Dimension		Position
	a	b	(x, y)
A	20 λ	10 λ	(10 λ, 15 λ)
B	10 λ	10 λ	(10 λ, -10 λ)
C	5 λ	20 λ	(-20 λ, 7.5 λ)

Table 2. Objects and scene configuration

		Objects		
	Radius	A	B	C
MMSE	$\widehat{a}$	19.93 λ	10.28 λ	5.58 λ
	$\widehat{b}$	9.44 λ	10.64 λ	21.02 λ
AMF	$\widehat{a}$	19.59 λ	10.04 λ	5.58 λ
	$\widehat{b}$	9.92 λ	10.16 λ	21.02 λ

Table 3. MMSE/AMF objects radius estimation; SNR = 5 dB

In spite of reduced range of model's validity near the forward scattering region, the parameter estimation procedure works fine and with this approach the geometrical parameters of elliptical cylinders are estimated.

As a future development of this work, the electromagnetic interaction between targets will be taken into account and we will investigate the possibility of using this model in the reconstruction methods. Although the objects with circular and elliptical cross sections are common in the nature, as trees, other geometrical forms, like triangular, are not studied here and the influence of diffraction by corners could also be taken in account in future work.

References

[1] J. C. Stover, *Optical Scattering, Measurement and Analysis.* Bellingharn, Washington. Second edition, 1995.
[2] R. Mittra, *Computer techniques for electromagnetics.* Pergamon, New York, 1973.
[3] G. L. Turin, "An introduction to matched filters," *IRE Transactions on Information Theory 6: 311- 329,* June 1960.
[4] M. Barkat, *Signal detection and estimation.* ARTECH HOUSE, Second edition, 2005.
[5] R. Harrington, *Time Harmonic Electromagnetic Fields.* McGraw-Hill, New York, 1961.
[6] M. Abramowitz and I. A. Stegun, *Handbook of Mathematical Functions with Formulas, Graphs, and Mathematical Tables.* New York: Dover, tenth gpo printing ed., 1964.
[7] N. Bleistein, "Physical optics far field inverse scattering in the time domain," *J. Acoust. Soc. Am. 60,1259-1264,* 1976.
[8] L. Mandel and E. Wolf, *Optical Coherence and Quantum Optics.* Cambridge: Cambridge University Press, 1995.

Electromagnetic time reversal and scattering by a small dielectric inclusion

Souhir Gdoura[1], Abdul Wahab[2] and Dominique Lesselier[1]

[1]Département de Recherche en Electromagnétisme, Laboratoire des Signaux et Systèmes UMR8506, 91192 Gif-sur-Yvette cedex, France

[2]Centre de Mathématiques Appliquées, UMR 7641, École Polytechnique, 91128 Palaiseau, France

E-mail: souhir.gdoura@lss.supelec.fr, wahab@cmap.polytechnique.fr, dominique.lesselier@lss.supelec.fr

Abstract. An asymptotic expression of the scattered electromagnetic far-field due to a small spherical dielectric inclusion illuminated by an impulsive vector source nearby is proposed first. Then, a time-reversal technique enabling to locate the position of this inclusion is developed and its performances are illustrated from numerical simulations.

1. Introduction

In previous works, e.g., [4], asymptotic formulas of the electromagnetic fields scattered by a small inclusion (dielectric, magnetic, or perfectly conducting as well) or a finite collection of them have been developed in the frequency domain up to leading order vs. frequency and size of the inclusion. In acoustics and elasticity, similar asymptotic expansions have been rigorously established for both near- and far-fields scattered by similar inclusions in the frequency domain, again at leading order. From these expansions, one can investigate transient-wave cases by truncating (to summarize) the high frequencies and using inverse Fourier transforms, [1, 2]. This approach has been investigated and numerical simulations proposed in [9].

In this contribution, the electromagnetic case is investigated. A small 3-D bounded inclusion D with ideal dielectric permittivity ϵ and air magnetic permeability μ_0 (no magnetic behavior is set forth but the analysis extends to it) is considered, letting $D = \delta B + \mathbf{z}$, where B is a regular enough bounded domain in $\mathbb{R}^3$ which is representing the volume of the inclusion, $\mathbf{z}$ is the vector position of its center, and δ is the scale factor. This inclusion is located in the background (air) medium with air permittivity ϵ_0, letting $c = 1/\sqrt{\mu_0\epsilon_0}$ as the speed of light. It is illuminated by a point electric dipole at location $\bar{\mathbf{y}}$ assumed far away from $\mathbf{z}$. The question is threefold: Can one get a meaningful asymptotic expression of the scattered magnetic far-field in the time-harmonic regime? How does the corresponding formulation develop in the time-domain? Is time reversal as an imaging method a good mean to locate the inclusion?

The organization of the contribution is in parallel to the questioning. First, the asymptotic formula of the scattered magnetic far-field in frequency is introduced. Then, the one in time, by truncating high frequencies and applying the inverse Fourier transform, is proposed. Finally, time-reversal is developed and numerically illustrated as a method of localization of the inclusion.

2. Asymptotic formula of the scattered magnetic field in frequency domain

As already said, an ideal electric dipole with $\mathbf{e}$ vector direction is set at $\bar{\mathbf{y}}$ in air, and it is taken as the source of the illumination. One defines $\mathbf{J}_e = \mathbf{e}\delta(\mathbf{r}-\bar{\mathbf{y}})$ as the electric current density in space. At circular frequency ω (time-harmonic dependence as $+j\omega t$), the time-harmonic scattering problem in the presence of the inclusion reads as

$$\begin{cases} \nabla\times\left(\dfrac{1}{\epsilon_0}\nabla\times\mathbf{H}\right)-\omega^2\mu_0\mathbf{H}=\nabla\times\dfrac{1}{\epsilon_0}\mathbf{J}_e & \text{in } \mathbb{R}^3\setminus D \\ \nabla\times\left(\dfrac{1}{\epsilon}\nabla\times\mathbf{H}\right)-\omega^2\mu_0\mathbf{H}=0 & \text{in } \bar{D} \\ \dfrac{1}{\epsilon_0}(\nabla\times\mathbf{H})^+\times\nu-\dfrac{1}{\epsilon}(\nabla\times\mathbf{H})^-\times\nu=0 & \text{on } \partial D \\ \mu_0\mathbf{H}^+\cdot\nu-\mu_0\mathbf{H}^-\cdot\nu=0 & \text{on } \partial D \end{cases}$$

Here ν is the unit normal on ∂D, exponents $^+$ and $^-$ referring to limit values from the outside and the inside as usual. $\underline{\mathbf{G}}^{me}$ is the magnetic-electric dyadic Green function in free space, satisfying the dyadic Helmholtz equation

$$\nabla\times\nabla\times\underline{\mathbf{G}}^{me}-\frac{\omega^2}{c^2}\underline{\mathbf{G}}^{me}=\nabla\times\underline{\mathbf{I}}\delta\left(\mathbf{r}\right). \tag{1}$$

with proper behavior to be accounted for at infinity. $\mathbf{H}_0$ is the incident magnetic field (the field due to the dipole in the absence of the inclusion) such as

$$\nabla\times\left(\frac{1}{\epsilon_0}\nabla\times\mathbf{H}_0\right)-\omega^2\mu_0\mathbf{H}_0=\nabla\times\tfrac{1}{\epsilon_0}\mathbf{J}_e \quad \text{in } \mathbb{R}^3, \quad \text{and} \quad \nabla\cdot\mathbf{H}_0=0 \quad \text{in } \mathbb{R}^3$$

or equivalently $\mathbf{H}_0(\mathbf{r},\bar{\mathbf{y}},\omega)=-\underline{\mathbf{G}}^{me}(\mathbf{r},\bar{\mathbf{y}},\omega)\cdot\mathbf{e}$, where $\underline{\mathbf{G}}^{me}(\mathbf{r},\bar{\mathbf{y}})=\nabla\times\underline{\mathbf{I}}\dfrac{e^{-i\frac{\omega}{c}|\mathbf{r}-\bar{\mathbf{y}}|}}{4\pi|\mathbf{r}-\bar{\mathbf{y}}|}$, using the notation $(\mathbf{r},\bar{\mathbf{y}},\omega)$ to mark the dependence upon the source dipole and frequency.

Let henceforth assume that the source point $\bar{\mathbf{y}}$ lies far enough from the inclusion, and that this is true as well for the observation point $\mathbf{x}$. One derives an asymptotic representation of the magnetic field in the frequency domain when $(\omega/c)\delta << 1$ [5]:

$$\left(\mathbf{H}-\mathbf{H}_0\right)\left(\mathbf{x},\bar{\mathbf{y}},\omega\right)\approx\delta^3\underline{\mathbf{G}}^{me}\left(\mathbf{x},\mathbf{z},\omega\right)\cdot\underline{\mathbf{M}}\left(\epsilon,B\right)\nabla\times\mathbf{H}_0\left(\mathbf{z},\bar{\mathbf{y}},\omega\right), \tag{2}$$

wherein the (frequency-independent) polarization tensor $\underline{\mathbf{M}}\left(\epsilon,B\right)$ in the case of a small dielectric spherical inclusion reduces to [3] $\underline{\mathbf{M}}\left(\epsilon,B\right)=3(\epsilon-\epsilon_0)(\epsilon+2\epsilon_0)^{-1}|B|\underline{\mathbf{I}}$, with immediate extensions to the general triaxial ellipsoid and oblate and prolate spheroids.

The asymptotic formula is given at leading order in terms of frequency and size of the inclusion. The latter should be small compared to the wavelength of the wave impinging upon it. Then, as known, it behaves as a radiative electric dipole.

3. Asymptotic formula of the scattered magnetic field in time domain

Now, one considers the time-domain counterpart of the previous results. The magnetic-electric dyadic Green function $\underline{\mathbf{G}}^{me}$ is solution of the dyadic Helmholtz equation

$$\nabla\times\nabla\times\underline{\mathbf{G}}^{me}+\frac{1}{c^2}\frac{\partial^2}{\partial t^2}\underline{\mathbf{G}}^{me}=\nabla\times\underline{\mathbf{I}}\delta\left(\mathbf{r}\right)\delta\left(t\right) \tag{3}$$

with again proper behavior at infinity. One easily shows that

$$\underline{\mathbf{G}}^{me}\left(\mathbf{r},t\right)=\nabla\times\underline{\mathbf{I}}\frac{\delta\left(t-|\mathbf{r}|/c\right)}{4\pi|\mathbf{r}|}=\left[\frac{\delta\left(t-|\mathbf{r}|/c\right)}{|\mathbf{r}|}+\frac{\delta'\left(t-|\mathbf{r}|/c\right)}{c}\right]\frac{\hat{\mathbf{r}}\times\underline{\mathbf{I}}}{4\pi|\mathbf{r}|} \tag{4}$$

where $\hat{\mathbf{r}} = \mathbf{r}/|\mathbf{r}|$, $\delta'(\alpha) = d\delta(\alpha)/d(\alpha)$. As for the incident magnetic field radiated by an electric dipole source at $\bar{\mathbf{y}}$, it is given in the time domain by [8]:

$$\mathbf{H}_0(\mathbf{r}, \bar{\mathbf{y}}, t) = -\underline{\mathbf{G}}^{me}(\mathbf{r}, \bar{\mathbf{y}}, t) \cdot \mathbf{e}. \tag{5}$$

To derive the time-domain asymptotic formula, one needs to truncate the high frequencies. Once those higher than ω_c are taken off, the Dirac impulsion exciting the electric dipole becomes

$$\psi(t) = \int_{|\omega|\leq\omega_c} e^{j\omega t} \frac{d\omega}{2\pi} = \frac{\sin \omega_c t}{\pi t}. \tag{6}$$

The operator P_{ω_c} can be defined as a tempered distribution by $P_{\omega_c}[\psi](t) = \int_{|\omega|\leq\omega_c} \hat{\psi}(\omega) e^{j\omega t} \frac{d\omega}{2\pi}$. Applying the inverse Fourier transform of (2) after truncating the high frequencies, a truncated equation satisfied by the incident field follows as

$$\begin{cases} \nabla \times \left(\frac{1}{\epsilon_0} \nabla \times P_{\omega_c}[\mathbf{H}_0] \right) + \mu_0 \frac{\partial^2}{\partial t^2} P_{\omega_c}[\mathbf{H}_0] = \nabla \times \delta(\mathbf{r} - \bar{\mathbf{y}}) \psi(t) \mathbf{e} & \text{in } \mathbb{R}^3 \times \mathbb{R} \\ \nabla \cdot P_{\omega_c}[\mathbf{H}_0] = 0 \quad \text{in } \mathbb{R}^3 \times \mathbb{R} \end{cases}$$

If the point of observation is away from the inclusion, the asymptotic formula of the scattered magnetic field, when $\omega << c/\delta$, can be written as :

$$P_{\omega_c}[\mathbf{H} - \mathbf{H}_0](\mathbf{x}, \bar{\mathbf{y}}, t) \approx \delta^3 \int_{\mathbb{R}} P_{\omega_c}[\underline{\mathbf{G}}^{me}](\mathbf{x}, \mathbf{z}, t - \tau) \cdot \underline{\mathbf{M}}(\epsilon, B) \nabla \times P_{\omega_c}[\mathbf{H}_0](\mathbf{z}, \bar{\mathbf{y}}, \tau)\, d\tau, \tag{7}$$

In the above, the truncated dyadic Green function $P_{\omega_c}[\underline{\mathbf{G}}^{me}]$ is

$$P_{\omega_c}[\underline{\mathbf{G}}^{me}](\mathbf{r}, t) = \left[\frac{\psi(t - |\mathbf{r}|/c)}{|\mathbf{r}|} + \frac{\psi'(t - |\mathbf{r}|/c)}{c} \right] \frac{\hat{\mathbf{r}} \times \underline{\mathbf{I}}}{4\pi|\mathbf{r}|}. \tag{8}$$

From (7), scattering by the inclusion is equivalent to radiation of an electric dipole with specific time-dependence.

The *sinc* signal is not causal. To avoid this non-causality problem, one could instead use as source pulse on the electric dipole (this will impact the incident field and thus the scattered field) a time-shifted *sinc* centred at large enough t_s so as to be close to zero at the time origin (computationally speaking, several orders of magnitude smaller than the peak value). One could also introduce a Gaussian pulse and shift it. A possibly better solution would be to use as source pulse the derivative of the Blackman-Harris one [10], to maintain causality while ensuring a reasonably wide band, truncation then being effected at ω_c large vs. the usable bandwidth.

4. Time reversal

The idea of time-reversal is to record the transient tangential components of the wavefield on a closed surface surrounding the inclusion, and to retransmit them onto the same background in time-reversed chronology. Then the wave should refocus at the location of the inclusion[7, 6].

So, one assumes that the tangential components of the fields are collected on a large sphere S with normal ν surrounding the inclusion D for a sufficiently large window of time t_0. Time-reversal is described by the transform $t \mapsto t_0 - t$. Then, both tangential components on S are time-reversed and emitted from S. A time-reversed field $\mathbf{H}_{\text{tr}}$ propagates into the interior volume. Using the asymptotic formula (7), one can prove that it is approximated by [2]

$$\begin{aligned} \mathbf{H}_{\text{tr}}(\mathbf{x}, \bar{\mathbf{y}}, t) \approx & \int_{\mathbb{R}} dt' \int_S \left[\underline{\mathbf{G}}^{ee}(\mathbf{x}, \mathbf{x}', t - t') \cdot (\nu \times \nabla \times P_{\omega_c}[\mathbf{H} - \mathbf{H}_0](\mathbf{x}', \bar{\mathbf{y}}, t_0 - t')) \right. \\ & \left. + \nabla \times \underline{\mathbf{G}}^{ee}(\mathbf{x}, \mathbf{x}', t - t') \cdot (\nu \times P_{\omega_c}[\mathbf{H} - \mathbf{H}_0](\mathbf{x}', \bar{\mathbf{y}}, t_0 - t')) \right] d\sigma(\mathbf{x}'). \end{aligned} \tag{9}$$

the dependence upon $\mathbf{z}$ in the reversed field being implied, where $\underline{\mathbf{G}}^{ee}$ is defined in [11].

One gets

$$\begin{aligned}\mathbf{H}_{\text{tr}}(\mathbf{x},\bar{\mathbf{y}},t) \approx& -\delta^3 \int_{\mathbb{R}} d\tau \int_{\mathbb{R}} dt' \int_S \left[\underline{\mathbf{G}}^{ee}(\mathbf{x},\mathbf{x}',t-t')\cdot\left(\nu\times\nabla\times P_{\omega_c}\left[\underline{\mathbf{G}}^{me}\right](\mathbf{x}',\mathbf{z},t_0-\tau-t')\right)+\right.\\ &\left.\nabla\times\underline{\mathbf{G}}^{ee}(\mathbf{x},\mathbf{x}',t-s)\cdot\left(\nu\times P_{\omega_c}\left[\underline{\mathbf{G}}^{me}\right](\mathbf{x}',\mathbf{z},t_0-\tau-t')\right)\right]\cdot\mathbf{p}(\mathbf{z},\bar{\mathbf{y}},\tau)d\sigma(\mathbf{x}').\end{aligned}$$

Since

$$\begin{aligned}&\int_{\mathbb{R}} dt' \int_S \left[\underline{\mathbf{G}}^{ee}(\mathbf{x},\mathbf{x}',t-t')\cdot\left(\nu\times\nabla\times P_{\omega_c}\left[\underline{\mathbf{G}}^{me}(\mathbf{x}',\mathbf{z},t_0-\tau-t')\right)\right]+\right.\\ &\left.\nabla\times\underline{\mathbf{G}}^{ee}(\mathbf{x},\mathbf{x}',t-t')\cdot\left(\nu\times P_{\omega_c}\left[\underline{\mathbf{G}}^{me}\right](\mathbf{x}',\mathbf{z},t_0-\tau-t')\right)\right]d\sigma(\mathbf{x}')\\ &=\mu_0\frac{\partial}{\partial t}\left[P_{\omega_c}\left[\underline{\mathbf{G}}^{me}(\mathbf{x},\mathbf{z},t_0-\tau-t)\right]-P_{\omega_c}\left[\underline{\mathbf{G}}^{me}(\mathbf{x},\mathbf{z},t-t_0+\tau)\right]\right],\end{aligned}\tag{10}$$

one arrives at

$$\begin{aligned}\mathbf{H}_{\text{tr}}(\mathbf{x},\bar{\mathbf{y}},t)\approx-\delta^3\int_{\mathbb{R}}\quad&\mu_0\frac{\partial}{\partial t}\left[P_{\omega_c}\left[\underline{\mathbf{G}}^{me}(\mathbf{x},\mathbf{z},t_0-\tau-t)\right]-P_{\omega_c}\left[\underline{\mathbf{G}}^{me}(\mathbf{x},\mathbf{z},t-t_0+\tau)\right]\right]\\ &\cdot\mathbf{p}(\mathbf{z},\bar{\mathbf{y}},\tau)\,d\tau,\end{aligned}\tag{11}$$

where $\mathbf{p}(\mathbf{z},\bar{\mathbf{y}},\tau)=\underline{\mathbf{M}}(\epsilon,B)\cdot\nabla\times P_{\omega_c}\left[\mathbf{H}_0\right](\mathbf{z},\bar{\mathbf{y}},\tau)$.

The interpretation is that there is superposition of ingoing and outgoing waves, centered at location $\mathbf{z}$ of the inclusion. To show it more clearly, one assumes that $\mathbf{p}(\mathbf{z},\bar{\mathbf{y}},\tau)$ is concentrated at $\tau=T=|\mathbf{z}-\bar{\mathbf{y}}|/c$, which is reasonable since $\mathbf{p}(\mathbf{z},\bar{\mathbf{y}},\tau)$ peaks at $\tau=T$. Then, (11) becomes

$$\mathbf{H}_{\text{tr}}(\mathbf{x},\bar{\mathbf{y}},t)\approx-\delta^3\mu_0\frac{\partial}{\partial t}\left[P_{\omega_c}\left[\underline{\mathbf{G}}^{me}(\mathbf{x},\mathbf{z},t_0-T-t)\right]-P_{\omega_c}\left[\underline{\mathbf{G}}^{me}(\mathbf{x},\mathbf{z},t-t_0+T)\right]\right]\cdot\mathbf{p}(\mathbf{z},\bar{\mathbf{y}},T),\tag{12}$$

indeed exhibiting ingoing and outgoing spherical waves.

Taking the Fourier transform of (10) and (11), letting $*$ denote the conjugate, one obtains:

$$\begin{aligned}\mathbf{H}_{\text{tr}}(\mathbf{x},\bar{\mathbf{y}},\omega)\approx&\int_S\left[\underline{\mathbf{G}}^{ee}(\mathbf{x},\mathbf{x}',\omega)\cdot\left(\nu\times\nabla\times\left[\mathbf{H}-\mathbf{H}_0\right]^*(\mathbf{x}',\mathbf{z},\omega)\right)\right.\\ &\left.+\nabla\times\underline{\mathbf{G}}^{ee}(\mathbf{x},\mathbf{x}',\omega)\cdot\left(\nu\times\left[\mathbf{H}-\mathbf{H}_0\right]^*(\mathbf{x}',\mathbf{z},\omega)\right)\right]d\sigma(\mathbf{x}')\\ =&-2\omega\mu_0\delta^3\Im m\underline{\mathbf{G}}^{me}(\mathbf{x},\mathbf{z},\omega)\cdot\underline{\mathbf{M}}(\epsilon,B)\nabla\times\mathbf{H}_0(\mathbf{z},\mathbf{y},\omega).\end{aligned}\tag{13}$$

The time-reversed field is proportional to the imaginary part of the Green dyad (manifesting the Rayleigh resolution, not withstanding polarization-dependent phenomena). Refer to classical results on the generalized Porter-Bojarski integral equation, e.g., among many, [12].

5. Numerical simulations

A dielectric inclusion of diameter $\lambda/25$ where $\lambda=0.75$m and relative permittivity 3 is placed at the origin of space. It is illuminated by a point dipole at $(6\lambda=4.5\text{ m},0,0)$ position (i.e., $T=15$ns), with unit direction $\mathbf{e}=[0,1,0]$. All frequencies above 400 MHz are truncated. Figure 1 shows the excitation signal *sinc* shifted by $t_s=20*(2\pi/\omega_c)=50$ns and the z-component of the scattered magnetic field computed by the asymptotic formula in the time domain as function of time (at the location of the source see Fig. 2). As expected, the scattered field is concentrated around the propagation time of the wave, at $t=2T+t_s=80$ns.

Time reversal is applied to locate the inclusion at instant $t=t_0-T$ in the plane (x,y), where $t_0=140$ns (one is safely close to a 0 magnitude of the recorded field). Figure 3 indeed depicts an anti-symmetric focal spot at the position of the inclusion.

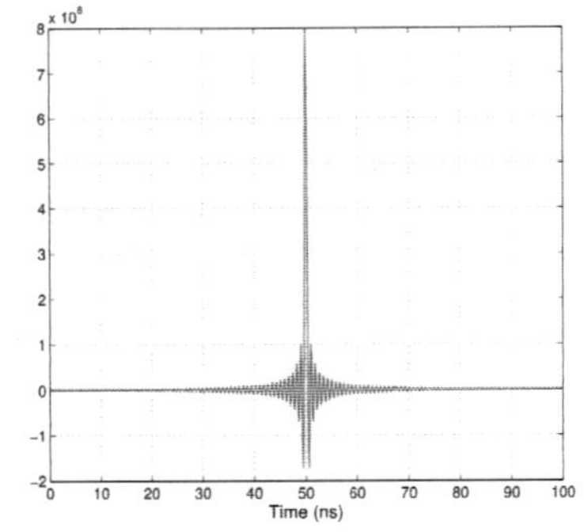

Figure 1. sinc excitation shifted by $t_s = 50$ns.

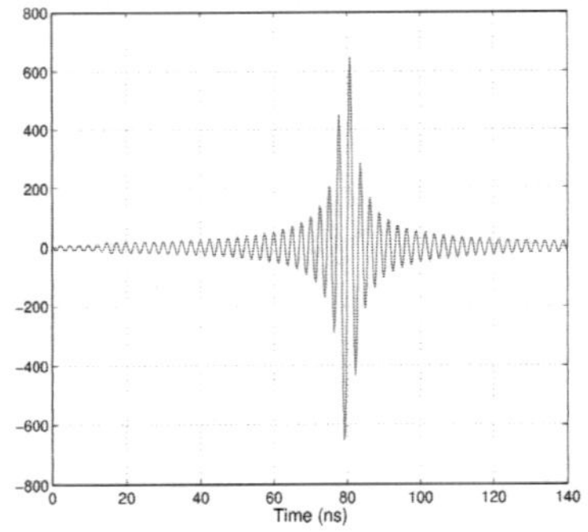

Figure 2. z-component of the scattered magnetic field.

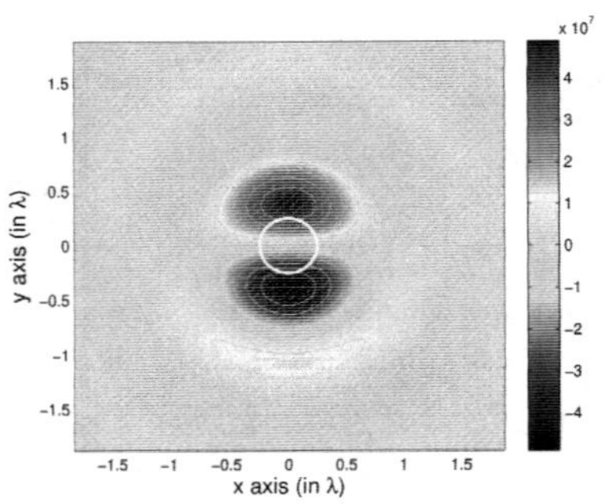

Figure 3. Time reversal of the z-component of the magnetic field in plane (x, y) at time $t = 125$ns.

Now, one excites the dipole by a causal signal as the first derivative of the Blackman-Harris pulse with central frequency $f_c = 100$ MHz where $\lambda_c = 3$m (to respect the bandwidth limitation), the duration of the source being $T_c = 1.55/f_c = 15.5$ns (see Fig. 4). Figure 5 shows that the z-component of the scattered field at the source location as a function of time is concentrated at instant $t = 37.75$ns with duration T_c .

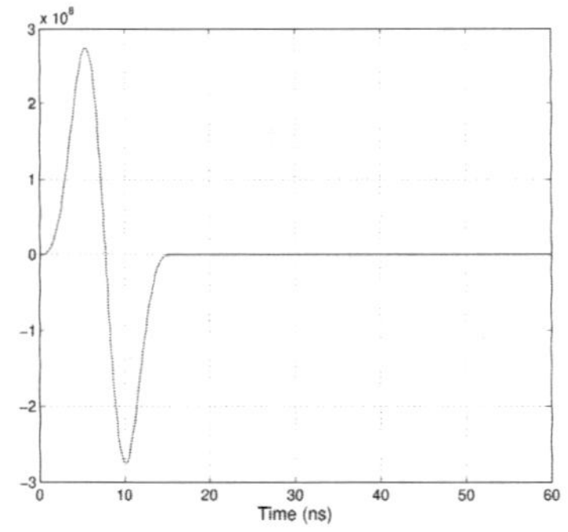

Figure 4. Derivative of Blackman-Harris signal excitation.

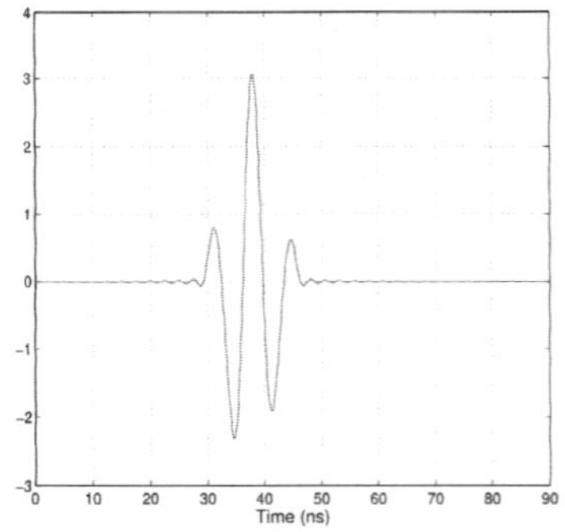

Figure 5. The z-component of the scattered magnetic field.

In figure 6, we plot the z-component of the time-reversed magnetic field at instant 67.25ns in the plane (x, y), letting $t_0 = 90$ns.

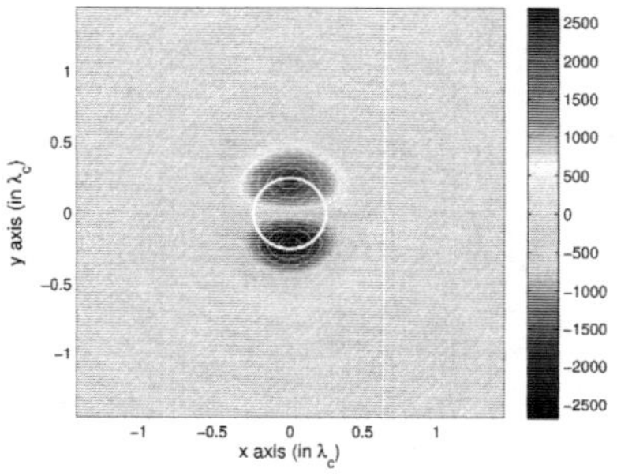

Figure 6. Time reversal of the z-component of the magnetic field in plane (x, y) at time $t = 67.25$ns.

6. Conclusion

Asymptotic formulas of the magnetic field (per duality, electric, if need be) scattered by a small dielectric inclusion in the time domain have been validated. Time-reversal enables to refocus to some extent onto the position of the inclusion. The next step is to study the effect of dispersion and attenuation of the background medium involved on the performance of this approach.

References

[1] H. Ammari *et al.* 2011 Transient elasticity imaging and time reversal Proc. Royal Society Edinburgh, Sect. A Mathematics 141, pp. 1121-1140.

[2] H. Ammari *et al.* 2010 Transient anomaly imaging by the acoustic radiation force, J. Differ. Equat. 249, pp. 1579–1595.

[3] H. Ammari and H. Kang 2007 *Polarization and Moment Tensors: with Applications to Inverse Problems and Effective Medium Theory*, Springer, New York.

[4] H. Ammari *et al.* 2007 MUSIC-type electromagnetic imaging of a collection of small 3-D bounded scatterers, SIAM J. Sci. Comput. 29, pp. 674-709.

[5] H. Ammari *et al.* 2001 Asymptotic formulas for perturbations in the electromagnetic fields due to the presence of inhomogeneities of small diameter II. The full Maxwell equations. Journal de Mathmatiques Pures et Appliques 80, pp. 769-814

[6] R. Carminati *et al.* 2007 Theory of the time reversal cavity for electromagnetic fields, Opt. Lett. 32 pp. 3107-3109.

[7] D. Cassereau *et al.* 1992 Time-reversal of ultrasonic fields. III. Theory of the closed time-reversal cavity, IEEE Trans. Ultrasonics, Ferroelect. Freq. Control 39, pp. 579–592.

[8] L. B. Felsen and N. Marcuvitz 1994 *Radiation and Scattering of Waves*, Wiley-IEEE p. 17.

[9] S. Gdoura and L. Guadarrama Bustos 2011 Transient wave imaging of anomalies: a numerical study, in *Mathematical and Statistical Methods for Imaging*. AMS series 548, H. Ammari *et al.* eds.

[10] M. E. Yavuz *et al.* 2006 Full time-domain DORT for ultrawideband electromagnetic fields in dispersive, random inhomogeneous media, IEEE Trans. Antennas Propagat., 54, pp. 2305–2315.

[11] A. D. Yaghjian and T. Hansen 1999 *Plane-Wave Theory of Time-Domain Fields: Near-Field Scanning Applications*, IEEE Press, p.182.

[12] R. P. Porter and A. J. Devaney 1982 Holography and the inverse source problem, J. Opt. Soc. Am. 72, pp. 327–330.

2nd Int. Workshop on New Computational Methods for Inverse Problems (NCMIP 2012) IOP Publishing
Journal of Physics: Conference Series **386** (2012) 012011 doi:10.1088/1742-6596/386/1/012011

A Level-Set Approach for Radar Imaging From Ramp Response in Arbitrary Directions

Y Wen[1], N de Beaucoudrey[1], J Saillard[1], J Chauveau[2] and P Pouliguen[3]
[1] IETR UMR 6164 CNRS, Université de Nantes, Lunam Université, Nantes, France
[2] DGA Maîtrise de l'Information, Bruz, France
[3] DGA DS/MRIS, Bagneux, France

E-mail: yan-hua.wen@etu.univ-nantes.fr, nicole.de-beaucoudrey@univ-nantes.fr

Abstract. Low frequency Radar imaging can be used to reconstruct the global shape of targets using the ramp response technique, which only needs no more than 3 observing angles. We have developed a new algorithm permitting to generate promising shapes of targets for mutually orthogonal directions, but giving distorted results otherwise. To solve this ill-posed inverse problem, we use the level set method, which iteratively deforms the shape of the target under a velocity field, and we analyze the property of the desirable velocity. Numerical results obtained by this method in arbitrary directions are given as well.

1. Introduction

Among inverse scattering methods for radar imaging, the ramp response technique [1] only needs no more than 3 observing angles. The ramp response, which is the transient scattering response from an incident wave in the form of ramp, is related to the cross-sectional area of the target perpendicular to the observing direction. This area, named profile function, can be used to reconstruct the global shape of the target. The reconstruction algorithm originally proposed by Young [2] has been applied for years to image scattering [3-5], as well as to underwater acoustic imaging [6]. This algorithm is limited to convex and single objects, that is why we developed a new algorithm [7] able to overcome this limitation. Both algorithms work well for mutually orthogonal directions, while they produce distorted shapes of the target in non-orthogonal cases.

Therefore, an optimization procedure is required to evolve towards a promising estimate by minimizing the cost functional. The level set method, devised by Osher and Sethian [8] [9], represents the evolving object as the zero level of an implicit higher dimensional function. This allows topology changes, such as splitting and merging of connected components during the deformation of the object, in a completely automatic and implicit manner. This advantage is of great importance for inverse problem since it requires no a priori assumption about the object geometry. It has been proved to be effective in retrieving separated objects from a single initial guess and has been illustrated in the inverse scattering problem [10-12]. Consequently, we apply this method to obtain a satisfactory shape of the target.

2. Image reconstruction from profile functions

The ramp response of a target, $h_r(t)$, is defined by the far zone transient scattered wave resulting from illumination by a plane electromagnetic wave with a temporal ramp wave shape. In time domain, it is the second integral of its impulse response, $h_i(t)$ and it can be expressed as the Inverse Fourier Transform (IFT) of the weighted transfer function, $H_r(j\omega)$:

$$h_r(t) = \int_{-\infty}^{t}\int_{-\infty}^{t'} h_i(t'')dt''dt' = IFT[H_r(j\omega)] = IFT\left[\frac{H(j\omega)}{(j\omega)^2}\right] \tag{1}$$

The monostatic ramp response is directly related to the profile function of the target, $A(u)$, which is the transverse cross-sectional area of the target as a function of the distance along the line-of-sight [1]

$$h_r(t) \approx -\frac{1}{\pi c^2} A(u) \quad \text{with} \quad u = \frac{ct}{2} \tag{2}$$

where c is the speed of light in free space, t the time variable, and u the space variable.

Using as target the example of figure 1(a), two types of profile functions (PF) are compared in figure 1(b): 1/ the ideal PF, "geometrical", equal to the cross-sectional area $A(u)$, for $u = x,y,z$; 2/ the real PF, "physical", calculated by (2) from the ramp response. To ensure that the physical PF is a valid estimate of the geometrical PF, it is necessary to match the frequency band to target dimensions, choosing the upper Rayleigh region and the resonance region of the target.

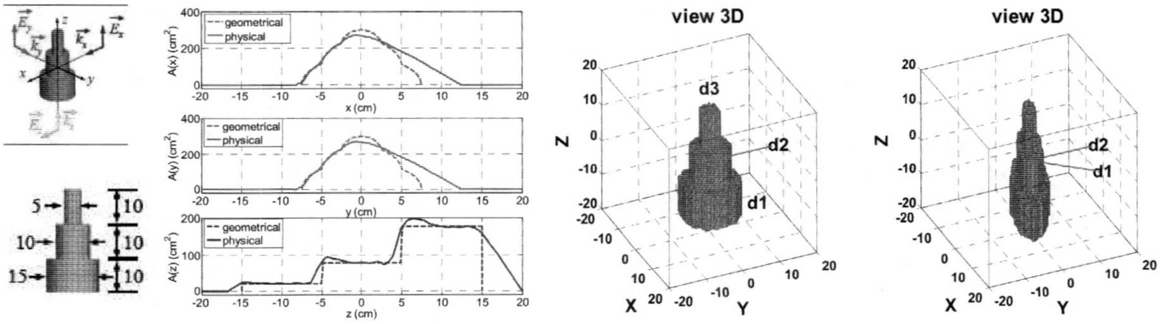

(a) (b) (c) (d)

Figure 1. (a) Configuration of a radar target; (b) geometrical and physical profile functions in 3 orthogonal directions: x, y, z; (c) reconstruction with 3 mutually orthogonal directions; (d) reconstruction with arbitrary directions.

The original algorithm, proposed by Young [2] for 3D reconstruction from ramp responses, is limited to convex and single object. We have therefore developed a new algorithm [7], which is able to overcome this limitation by exploiting more effectively the information contained in the profile functions. With 3 geometrical profile functions of the object in figure 1(b), the resulting reconstruction with mutually orthogonal directions, figure 1(c), shows a high similarity with the original object. Considering now the case of arbitrary directions (with one direction at 50° in plane xoy, and the other two directions unchanged along y and z), the reconstructed result is distorted, figure 1(d).

3. Optimization using the level-set method

To improve the performance of this algorithm for arbitrary directions, it is necessary to use an iterative process to obtain an optimal estimate of the target by minimizing the mismatch between the data, i.e. profile functions of the unknown object, and the profile functions of the updated object.
Our problem can be formalized in the form of

$$A^d u = g^d \tag{3}$$

where $g^d(M,1)$ is the observed data, the PF in direction d, $u(N,1)$ the vector representing the unknown binary object we are looking for, $A^d(M, N)$ the observing matrix (or mapping matrix) and d the index number of direction. d=1, 2, 3 in our case with three observing directions.

3.1. The forward problem

The forward problem, at step *k*, consists in calculating the profile functions of the evolving object with the following equation:

$$g_k^d(i) = \sum_{j=1}^{N} A^d(i,j)\, u_k(j) \quad \text{with} \quad i = 1, 2 \cdots M \tag{4}$$

For this, we developed an algorithm to calculate the profile function of a 3D object in an arbitrary direction: its principle is to distribute the pixels of the object in successive cutting planes perpendicular to the chosen direction. With the definition of profile functions, $A^d(i,j)$ represents the contribution that the pixel $u(j)$ gives to the slice i, and the profile function is obtained by summing individual contributions from all pixels. At iteration k, we define the cost functional, F_k, as the norm of the error between the PF calculated from the evolving object, g_k, and the PF observed from the real object, g:

$$F_k = \sum_d \frac{1}{2} \left\| g_k^d - g^d \right\|^2 \quad \text{with} \quad d = 1, 2, 3 \tag{5}$$

Applying the proposed algorithm, the three geometrical profile functions are compared in figure 2 with those calculated from the initial guess reconstructed from orthogonal (figure 1(c)) as well as arbitrary directions (figure 1(d)). In the orthogonal case, the cost function is negligible ($F_0 = 0.02$), whereas it is much more significant in the non-orthogonal case ($F_0 = 0.16$). Therefore, the cost functional behaves as a quantitative indicator for the evolution and quality evaluation of the estimate.

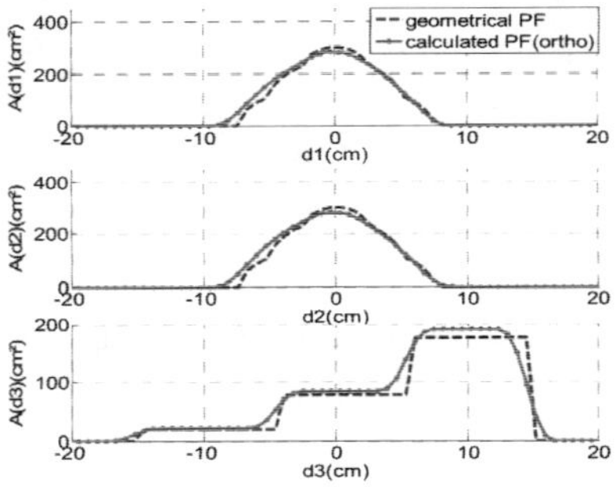

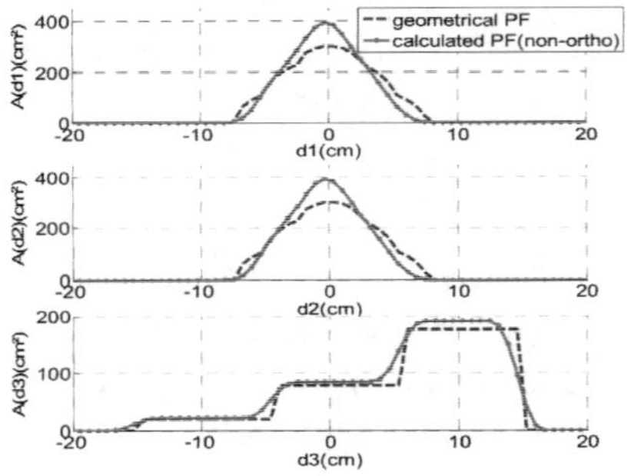

Figure 2. Comparison between the 3 geometrical PF with those calculated from the reconstruction of figure 1(c) and (d): (a) orthogonal case; (b) non-orthogonal case.

3.2. The inverse problem

Most of iterative methods evolve the explicit function of object shape during the iteration. On the contrary, the level set method represents the shape as the zero level of a higher order level set function, which requires no a priori assumption on the object geometry or structure. In a computational domain D enclosing the object Ω, the level set function $\varphi(x)$ is defined as:

$$\begin{cases} \varphi(\boldsymbol{x}) > 0 & \text{if} \quad \boldsymbol{x} \notin \Omega \\ \varphi(\boldsymbol{x}) < 0 & \text{if} \quad \boldsymbol{x} \in \Omega \\ \varphi(\boldsymbol{x}) = 0 & \text{if} \quad \boldsymbol{x} \in C \end{cases} \tag{6}$$

where $\boldsymbol{x}$ is the position variable, (x, y) and (x, y, z) for 2D and 3D case respectively, and C is the contour of the object. The deformation of the object is formalized as a Hamilton-Jacobi equation for the level set function φ under a normal velocity field V [8]:

$$\frac{\partial \varphi}{\partial t} + V\left|\nabla \varphi\right| = 0 \tag{7}$$

Following the formulation in [13], a common choice of velocity is:

$$V = -J(u)^T F(u)(u_i - u_o) \tag{8}$$

where $J(u)$ is the Jacobian of $F(u)$ at u. u_i *and* u_o are the prescribed values to represent the inside and outside of the desired unknown. In our case, $u_o = 0$ and $u_i = 1$ represent the binary unknown.

Shape deformation, 'expand' or 'shrink', is directly related to the change of the level set function, which is mainly determined by the sign of the velocity. From (7), when V is positive, φ decreases, therefore the shape expands in the normal direction of the contour; whereas, when V is negative, the shape shrinks. In the computational domain, as compared with the real object, the evolving object might have "correct pixels", P_c, "missing" pixels, P_m, which belong to the real object but are not selected by the evolution, and "false" pixels, P_f, which are in the opposite situation. The aim of the shape optimization is to iteratively fill up P_m and remove P_f. These two "actions" are directly related to the change of the level set function, which is determined by the sign of the velocity. Therefore, an effective velocity should have the properties shown in Table 1, where the subscript 'e' represents the evolving object and 'r' the real object. For P_m, the level set function φ should decrease, so V is positive, while for P_f, φ should increase, so V is negative.

Table 1. Properties of desirable velocity.

Pixel	P_m	P_f
u_e	0	1
u_r	1	0
φ_e	>0	<0
φ_r	<0	>0
φ	↓	↑
V	>0	<0

An example is given to explain how the velocity works during the evolution. Figure 3(a) compares the real object (red part) and the estimate (blue part) in a cut-plane of the computational domain. Therefore, there are some "missing" pixels, P_m, and "false" pixels, P_f, (figure 3(b)). According to the property of desirable velocity in table 1, figure 3(c) shows that the velocity is positive for P_m and negative for P_f. Therefore, after one iteration, all "false" pixels are removed well and some "missing" pixels vanish (figure 3(d)). Remaining "missing" pixels will be corrected after a few iterations with the positive velocity.

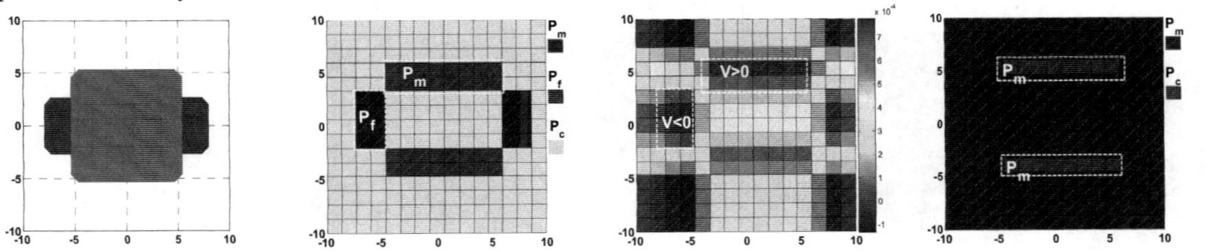

(a) (b) (c) (d)

Figure 3. (a) Comparison between the real object and the estimate; (b) the initial cut-plane; (c) the velocity for each pixel in this cut-plane; (d) the cut-plane after one iteration.

4. Numerical results

Now we consider to reconstruct a PEC cube (figure 4(a)) with the level set method. The computational domain is distributed into $N=16^3$ cells, with 16^2 pixels in each cut plane. To apply this method, an initial guess is required. It can be arbitrary, and we choose here a rectangular box to test the performance or this method, since it needs either to "expand" or to "shrink" for obtaining the shape of the original cube. Once the initial contour C_0 is given, the initial level set function φ_0 is defined as the signed distance between each point and C_0, negative inside and positive outside [10]. The gradient of the level set function, $|\nabla\varphi|$, is approximated by a Hamiltonian scheme [8]. Combining (5) with (8), the velocity can be easily obtained by the algorithm we proposed in the forward problem.

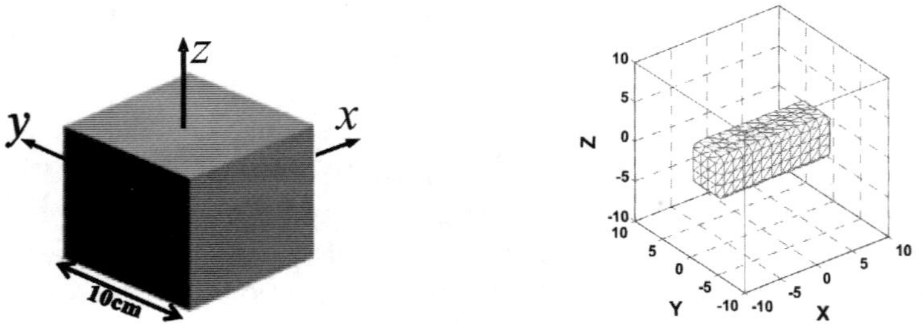

(a) (b)

Figure 4. (a) Configuration of a PEC cube; (b) the initial guess: a rectangular.

Here, we use the geometrical profile function as the observed data of the real object. With 3 mutually orthogonal directions and 20 iterations, the reconstructed result (figure 5) shows high similarity with some acceptable errors comparing to the real object. For the non-orthogonal case ($d1 = 45°$,y,z), the estimate (figure 6) has more "missing" and "false" pixels, but still it is not severely distorted. Now if we add another direction ($d4 = 60°$), the result (figure 7) shows greater agreement with the real shape. It is worth mentioning that more accurate images can be obtained with a larger number of samples.

5. Conclusion and perspectives

Radar imaging from ramp responses can reconstruct the shape of a target with only three directions. Our reconstruction algorithm overcomes the limitation of previous algorithms for orthogonal directions, but it produces distorted estimate in non-orthogonal case. The calculation of the profile function of a 3D object is used to compute the cost function for the evolution process. The level set method, which has been proved to be effective in shape optimization, is used. The property of an effective velocity for this method is analyzed. This method gives good results with 3 orthogonal directions and promising results with 4 non-orthogonal directions without restrictive requirement of the initial guess. Our further work is to construct an adapted cost functional so that the velocity satisfies desirable properties. The regularization of the evolution, for example uniqueness and convergence, should be considered as well.

References

[1] Kennaugh E M and Moffatt D L 1965 Transient and impulse response approximations *Proc IEEE* 53 893-901
[2] Young J D 1976 Radar imaging from ramp response signatures *IEEE Trans. Antennas Propagat.* AP-24 276-282
[3] Moffatt D L 1981 Ramp response radar imagery spectral content *IEEE Trans. Antennas Propagat.* 29 399-401
[4] Nag S and Peters L 2001 Radar images of penetrable targets generated from ramp profile functions *IEEE Trans. Antennas Propagat.* 49 32-40

Journal of Physics: Conference Series **386** (2012) 012011 doi:10.1088/1742-6596/386/1/012011

[5] Chen C and Peters L 2007 Radar scattering and target imaging obtained using ramp-response techniques *Antennas and Propagation Magazine* 49 13-27

[6] Zhang X Lib W and Liu G 2002 A new technique in ramp response for acoustic imaging of underwater objects *Applied Acoustics* 63 453-465

[7] Chauveau J and de Beaucoudrey N 2011 A new algorithm of 3D image reconstruction of radar targets from ramp responses in low frequency *PIER* M16 1-18

[8] Osher S and Sethian J 1988 A front propagating with curvature-dependent speed: algorithms based on Hamiton-Jacobi formulations *J. Comput. Phys.* 7 912-49

[9] Sethian J A 1999 *Level set methods and fast marching methods* (Cambridge: Cambridge University Press)

[10] Litman A, Lesselier D and Santosa F 1998 Reconstruction of a two-dimensional binary obstacle by controlled evolution of a level-set *Inverse Problems* 14 685-706

[11] Dorn O and Lesselier D 2006 Level set methods for inverse scattering *Inverse Problems* 22 R67–R131

[12] Hajihashemi M and El-Shenawee M 2011 Inverse scattering level set algorithm for retrieving the shape and location of multiple targets *Antennas and Propagation (EUCAP)* (Rome, Italy, 11-15 April *2011*) pp.3819-3823

[13] Santosa F 1996 A level set approach for inverse problems involving obstacles *Control Optimizat. Calculus Variat* 1 17-33

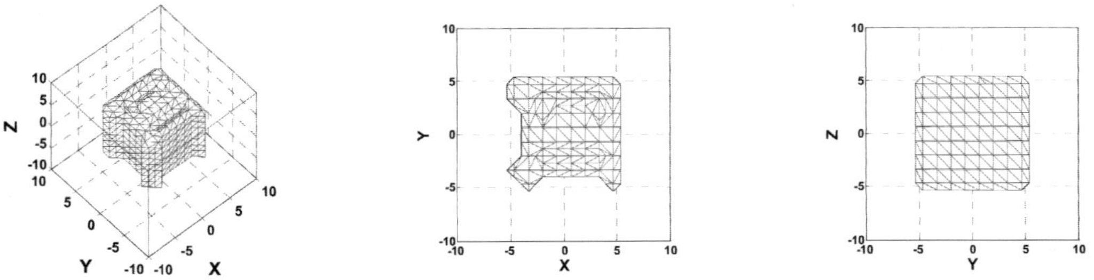

Figure 5. Result obtained with 3 mutually orthogonal directions after 20 iterations

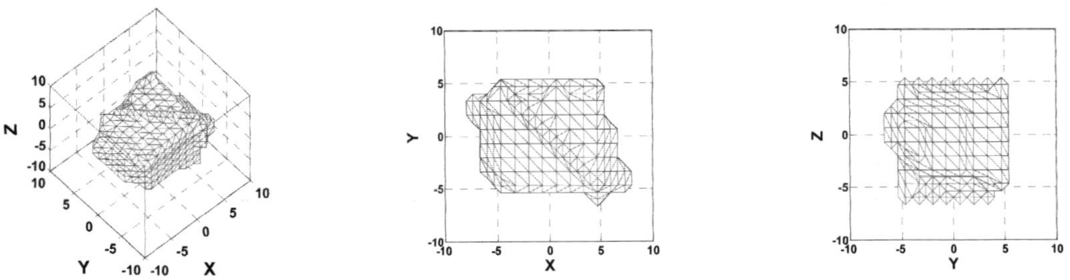

Figure 6. Result obtained with 3 non-orthogonal directions (d1=45°) after 20 iterations

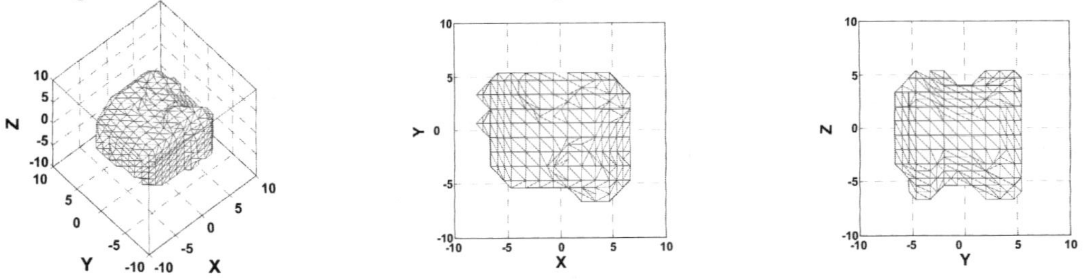

Figure 7. Result obtained with 4 non-orthogonal directions (d1=45°, d4=60°) after 20 iterations

Journal of Physics: Conference Series **386** (2012) 012012 doi:10.1088/1742-6596/386/1/012012

Absorption and phase retrieval in phase contrast imaging with non linear Tikhonov regularization

B Sixou[1], V Davidoiu[1], M Langer[1,2], and F Peyrin[1,2]

[1] Creatis, CNRS UMR 5220, Inserm U630, INSA de Lyon, Universite de Lyon, F-69621

[2] European Synchrotron Radiation Facility, 6 rue Jules Horovitz, F-38043, Grenoble Cedex France

E-mail: bruno.sixou@insa-lyon.fr

Abstract. The X-ray phase contrast imaging technique relies on the measurement of the Fresnel diffraction intensity patterns associated to a phase shift induced by the object. The simultaneous recovery of the phase and of the absorption is an ill-posed non linear inverse problem. In this work, we investigate the resolution of this problem with non linear Tikhonov regularization and with the Iterative Gauss Newton method. The algorithm is evalutated using simulated noisy data.

1. Introduction

X-ray microtomography is used in many applications such as bone imaging and material science [1]. Yet, it is difficult to image soft tissues, especially within dense structures. For coherent X-rays, phase contrast can be achieved by letting the beam propagate in the free space after interaction with the object [2, 3] with a drastic increase in sensitivity. The Fresnel intensity maps recorded set the inverse problem to retrieve the phase shift induced by the object. [1, 2, 3, 4, 5]. Several linear phase retrieval methods have already been investigated [1, 2, 3, 4, 5] and improved by non linear approaches recently [6]. Yet, all these approaches are based on a precise knowledge of the absorption. The simultaneous absorption and phase retrieval remains a challenge. This work investigates the resolution of the non linear ill-posed inverse problem associated to the simultaneous recovery of the absorption and of the phase. We successively present the direct problem of the image formation, the non-linear Tikhonov regularization of the inverse problem and results obtained from simulations on noisy data.

2. The direct problem of the image formation

Considering a thin object illuminated with coherent X-rays of wavelength λ, the interaction of X-rays with matter can be described by a transmittance function T of the coordinates X=(x,y) in a plane perpendicular to the propagation direction z of the X-rays:

$$T(X) = \exp[-B(X) + i\varphi(X)] = a(X)\exp[i\varphi(X)] \quad (1)$$

where $B(X)$ is the absorption and φ the phase induced by the object. The intensity detected at a distance D is given by the squared modulus of the exit wave:

$$I_D(X) = |T(X)) * P_D(X))|^2 \quad (2)$$

where $*$ denotes the 2D convolution of the transmitance with the Fresnel propagator:

$$P_D(X) = \frac{1}{i\lambda D} exp(i\frac{\pi}{\lambda D}|X)|^2) \tag{3}$$

The real and imaginary part of the refractive index can then be reconstructed by recording the intensity at different angles with tomographic reconstruction algorithms, when the phase shift $\varphi(X)$ and the absorption $B(X)$ are known. Most existing phase retrieval approaches are based on a linearization of the direct problem valid under some rather restrictive assumptions [1, 2, 3, 4, 5]. Recently, a non linear iterative approach based on the Frechet derivative of the intensity has improved the phase retrieval accuracy [6]. Yet, in this approach, the absorption $B(X)$, obtained from the intensity measured for D=0, is assumed to be known. When imaging at high resolution, it is very difficult to obtain an absorption image without phase contrast sufficiently close to the detector. And this motivates the study of the joint absorption and phase retrieval algorithms, leading to improved solutions for the phase $\phi(X)$ and the absorption $B(X)$.

3. The non linear inverse problem of the simultaneous absorption and phase recovery

In order to solve this non linear inverse problem, we have used a Tikhonov regularization method and the Iterative Gauss Newton method (IRGN).

3.1. Non linear Tikhonov regularization

In the following, we will consider that the phase and the absorption have a bounded support Ω and that the domain of the intensity operator $I_D(B,\varphi)$ belongs to the functional space $H \times H$ with $H = L^2(\Omega) \cap L^\infty(\Omega)$. The real intensity data will be denoted I and the noisy intensity I_δ. These intensities are assumed to be such that $\|I_\delta - I\|_{L_2(\Omega)} \leq \delta$. The aim of the non linear Tikhonov regularization is to minimize the functional T_α of the two components u=(B,φ):

$$T_\alpha(u) = \frac{1}{2}\|I_D(u) - I_\delta\|^2_{L_2(\Omega)} + \frac{\alpha}{2}\|u\|^2_{L_2(\Omega)} \tag{4}$$

where α is a regularization parameter. A minimizer u^δ_α always exists under the assumptions detailed in [7, 8]. If a minimum norm solution u^+ exists with $I_D(u^+)$=I, then, under these assumptions, for a sequence (I_δ), with $\|I_\delta - I\| \leq \delta$, an a priori regularization parameter choice $\alpha \sim \delta$, there exist constants C and E such that $\|u^+ - u^\delta_\alpha\| \leq C\sqrt{\delta}$ and $\|I_D(u^\delta_\alpha) - I\| \leq E\delta$. It will be shown in a forthcoming paper that $I_D(B,\varphi)$ satisfies all former assumptions.

3.2. The Iterated Gauss-Newton method

We have used the well-known Iterated Gauss-Newton method to minimize the regularization functional given in Eq.4. [9, 10, 11, 12]. Denoting $I_D(u_k)'$ the Fréchet derivative of the intensity at the current point, the discrepancy term in Eq.4 is replaced by its approximation at the current point u_k, $T_\alpha(u) = \frac{1}{2}\|I_D(u_k) - I_\delta + I_D(u_k)'(u - u_k)\|^2_{L_2(\Omega)}$ providing a strongly convex quadratic functional. The algorithm has been generalized by the introduction of a line search procedure with a variable step size τ_k [9, 10, 11, 12]. The update formula becomes:

$$u_{k+1} = u_k - \tau_k(I'_D(u_k))^* I'_D(u_k)) + \alpha I)^{-1}(I_D(u_k)^*(I_D(u_k) - y) + \alpha(u_k - u_*)) \tag{5}$$

where $F'(u_k)^*$ is the adjoint of the Fréchet derivative. The step length parameter τ_k is chosen in order to minimize the Tikhonov's functional along the descent direction with a backtracking strategy [9, 10, 11, 12]. The descent direction $d_k = (I'_D(u_k)^* I'_D(u_k)) + \alpha I)^{-1}(I_D(u^*_k)(I_D(u_k) - y) + \alpha(u_k - u_*))$ is obtained with the linear system $A_k d_k = -g_k$ from the symmetric positive definite

Journal of Physics: Conference Series **386** (2012) 012012 doi:10.1088/1742-6596/386/1/012012

matrix $A_k = I'_D(u_k)^* I'_D(u_k) + \alpha I$ and from the gradient $g_k = I_D(u_k)^*(I_D(u_k) - y) + \alpha(u_k - u_*)$. This system is solved iteratively by a conjugate gradient method. The stopping index of the iterations is k^* determined by the conditions, $\|y^\delta - I_D(u_{k^*}))\| \leq \delta < \|y^\delta - I_D(u_k))\|$ for $0 \leq k \leq k^*$. It is based on stopping when the residual is of the order of the data error [9, 10, 11, 12].

3.3. Explicit calculation of the Frechet derivative of the intensity

The Fréchet derivative of the operator I(B,φ) at the point (B_k, φ_k) is the linear operator $I'(B_k, \varphi_k)$ defined by the relation $I(B_k + \epsilon_1, \varphi_k + \epsilon_2) = I(B_k, \varphi_k) + I'(B_k, \varphi_k)(\epsilon_1, \epsilon_2) + O(\|(\epsilon_1, \epsilon_2)\|^2)$ It can be calculated explicitly with:

$$I'(B_k, \varphi_k)(\epsilon_1, \epsilon_2) = 2Real\{[(-\epsilon_1 - i\epsilon_2)(exp(-B_k - i\varphi_k)) * P_D]\}\{[exp(-B_k + i\varphi_k)] * \overline{P_D}\} \quad (6)$$

The adjoint $I'(B_k, \varphi_k)^* : H \to H \times H$ is then defined by $I'(B_k, \varphi_k)^*(u) = 2Real(v_1, v_2)$ with:

$$v_1 = \{[-u(exp(-B_k)exp(-i\varphi_k) * P_D)] * \overline{P_D}\}\{exp(-B_k)exp(i\varphi_k)\} \quad (7)$$

$$v_2 = \{[u(exp(-B_k)exp(-i\varphi_k) * P_D)] * \overline{P_D}\}\{iexp(-B_k)exp(i\varphi_k)\} \quad (8)$$

4. Results

4.1. Simulation of the diffracted intensity

Following [3], one phantom was defined with an absorption and a phase map displayed in Figures 1(a) and 1(b). This phantom consists of the projections of two spheres, each of homogeneous composition, the real and complex parts of the refractive index being proportional, with a different proportionality constant in each sphere. The absorption and phase values were discretized on a regular grid of size $N = 75 \times 75$ with a pixel size of 15 μm. Propagation in free space was simulated using Eq.2 for D=1.4 m and $\lambda = 0.5166\mathring{A}$. A Gaussian white noise with peak-to-peak signal-to-noise ratio (PPSNR) of 24 dB was added to the image. Our phase and absorption retrieval algorithms are not globally convergent algorithms. An approximate absorption map was first chosen, with increased absorption values and larger spheres. Then, the phase obtained with the linear Mixed approach presented in [3] has been used as the starting point of our simulations. The initial absorption and phase root mean square error are around 45%. This a priori guess of the solution ensures the convergence of the algorithms.

4.2. Results

Figure 2(a) and 2(b) display the normalized mean square error for the phase and the absorption as a function of the number of iterations, $\|\varphi^+ - \varphi\|_{L_2(\Omega)}/\|\varphi^+\|_{L_2(\Omega)}$ and $\|B^+ - B\|_{L_2(\Omega)}/\|B^+\|_{L_2(\Omega)}$. The final phase and absorption error maps are displayed in Figures 1(c) and 1(d) respectively. Some errors are still present and a local minimum is obtained but the errors on the phase and absorption have been significantly reduced with this algorithm. We have no proof that the source condition is satisfied for the absorption and phase map considered here. Yet, our results demonstrate the efficiency of the proposed algorithm.

5. Conclusion

In this work, a non linear iterative approach for the simultaneous phase and absorption retrieval has been proposed. This problem is ill-posed and thus Tikhonov regularization and the iterative Gauss Newton method are included to obtain stable solutions. The method was tested with simulated data in the presence of noise. Significant decrease of the phase and absorption root-mean-square errors are obtained. This method will be tested on experimental data and can be applied in various fields, such as biomedical science and material science.

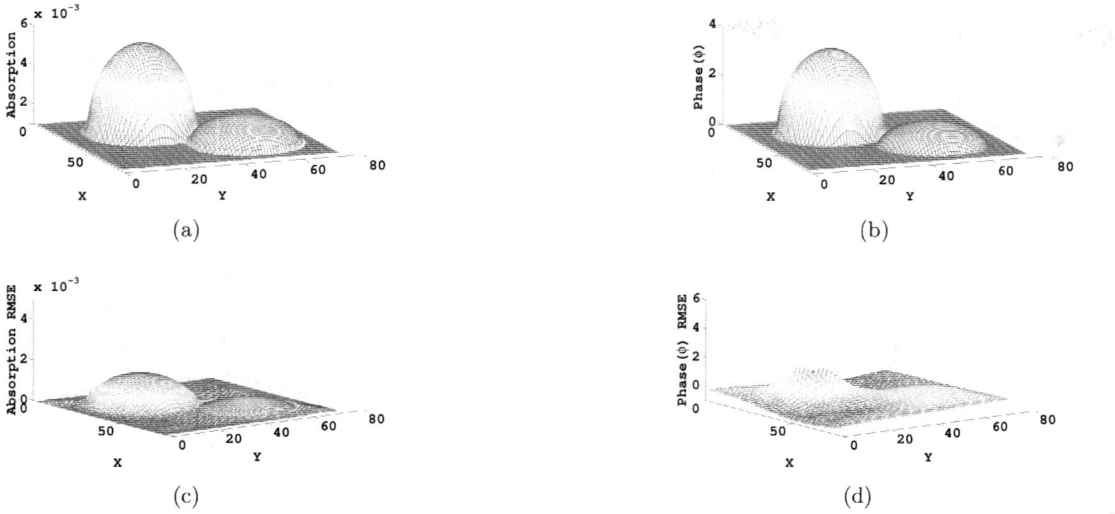

Figure 1. (a) Original absorption map, (b) Original phase map, (c) Absorption error map and (d) Phase error map at convergence.

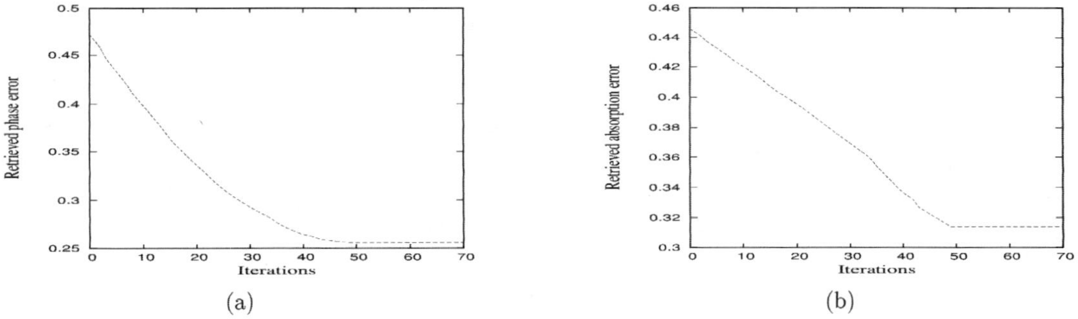

Figure 2. (a) Evolution of the phase RMSE with the iterations and (b) Evolution of the absorption RMSE with the iterations.

References

[1] Cloetens P, Barrett R, Baruchel J, Guigay J P, Schlenker M *J. Phys. D* 1996 **13** 133
[2] Paganin D M 2006 *Coherent X ray optics* (Oxford University Press, New York)
[3] Langer M, Cloetens P, Guigay J P, Peyrin F 2008 *Med. Phys* **35** 4556
[4] Gureyev T E, Nugent K A *J. Opt. Soc. Am. A* 1996 A **13** 1670
[5] Wilkins S W , Gureyev T E, Gao D, Pogany A, Stevenson A W *Phys. Rev. Lett* 1996 **77** 2961
[6] Davidoiu B, Sixou B, Langer M and Peyrin F 2011 *Opt. Express* **19** 22809
[7] Dicken V 1999 *Inverse Problems* **15** 931
[8] Engl H, Kunish K, Neubauer A 1989 *Inverse Problems* **5** 523
[9] Smirnoval A, Renaut A , Khan T 2007 *Inverse Problems* **23** 1547
[10] Bakushinsky A B 1992 *Comput. Math. Math. Phys* **32** 1353
[11] Blaschke B, Neubauer A, Scherzer O 1999 *IMA, Journal of Numerical Analysis* **17** 1997
[12] Bakushinsky A B, Smirnova A 2005 *Numer. Funct. Anal. Optm* **26** 35

Journal of Physics: Conference Series **386** (2012) 012013 doi:10.1088/1742-6596/386/1/012013

The truncated Newton method for Full Waveform Inversion

L. Métivier[1], R. Brossier[1], J. Virieux[1], S. Operto[2]

[1]ISTerre, Université de Grenoble, BP 53, 38041 Grenoble CEDEX 9, France, [2]Géoazur, La Darse, B.P. 48, 06235 Villefranche sur Mer CEDEX, France

E-mail: `ludovic.metivier@ujf-grenoble.fr`

Abstract. Full Waveform Inversion (FWI) is a promising seismic imaging method. It aims at computing quantitative estimates of the subsurface parameters (bulk wave velocity, shear wave velocity, rock density) from local measurements of the seismic wavefield. Based on a particular wave propagation engine for wavefield estimation, it consists in minimizing iteratively the distance between the predicted wavefield at the receivers and the recorded data. This amounts to solving a strongly nonlinear large scale inverse problem. This minimization is generally performed using gradient-based methods. We investigate the possibility of applying the truncated Newton (TrN) method to this problem. This is done through the development of general second-order adjoint state formulas that yield an efficient algorithm to compute Hessian-vector products, and the design of an adaptive stopping criterion for the inner conjugate gradient (CG) iterations. Numerical results demonstrate the interest of using the TrN method when multi-scattered waves dominate the recorded data.

1. Introduction

Full Waveform Inversion (FWI) computes quantitative estimates of the subsurface parameters, such as the bulk wave velocity, the shear wave velocity, or the rock density. Applications range from the localisation of natural resources, such as oil and gas, to reservoir and storage monitoring, and seismic risk prevention. The seismic data is collected through the so-called seismic experiment: several sources and receivers are located at the ground surface or in wells, and the wavefield generated by these sources is recorded locally by the receivers. Based on a wave propagation modeling (from the simplest acoustic equation to more sophisticated anisotropic visco-elastic dynamics), the FWI method consists in minimizing a misfit function, which measures a distance between the data predicted by the forward problem and the recorded data. We consider the general wave equation

$$S(p)u = \varphi, \tag{1}$$

where the subsurface parameters are denoted by $p \in \mathbb{R}^m$ (model space), the linear forward problem operator corresponding to the two-way wave equation [1] is denoted by $S(p)$, the source vector is denoted by φ, and the wavefield vector is denoted by u. These notations are general

[1] Note that the operator $S(p)$ depends non-linearly on parameter p

and can be applied either in the time domain or in the frequency domain. The FWI problem is expressed as the nonlinear least-square problem

$$\min_p f(p) = \frac{1}{2} \sum_{s=1}^{N_s} \|R_s u_s(p) - d_s\|^2, \tag{2}$$

where d_s and $u_s(p)$ are respectively the recorded dataset and the solution of the forward problem associated with the source φ_s, N_s is the total number of sources, and R_s is a restriction operator that maps the wavefield u_s to the receiver locations.

The minimization of $f(p)$ is based on the local Newton approach. A sequence p_k is computed from an initial guess p_0 using the update formula

$$p_{k+1} = p_k + \gamma_k \Delta p_k, \tag{3}$$

where the scalar γ_k is computed through a globalization method (linesearch, trust-region) and Δp_k is the solution of

$$\nabla^2 f(p_k) \Delta p_k = -\nabla f(p_k), \tag{4}$$

where the Hessian operator is denoted by $\nabla^2 f(p_k)$. The computation of the gradient $\nabla f(p)$ can be performed efficiently with the adjoint-state method [6]. Because of the large-scale aspect of the FWI problem (even for 2D applications, the problem often involves hundred thousands of unknowns parameters and discrete data), explicit computation and storage of $\nabla^2 f(p)$, *a fortiori* $(\nabla^2 f(p))^{-1}$, is prohibitive. Therefore, standard methods use an approximation Q_k of the inverse Hessian. Different choices can be made. For instance, choosing simply the identity matrix yields the steepest-descent method, which may converge slowly. A more sophisticated and efficient choice is the l-BFGS approximation [1]. This method estimates the effect of $\left(\nabla^2 f(p_k)\right)^{-1}$ on the gradient. At iteration k, this approximation is based on finite differences of the l previous values of the gradient.

However, Pratt et al [7] emphasize the crucial role of the inverse Hessian operator in FWI: its acts as a defocusing filter, improving the resolution of the subsurface parameter estimation, and can help to remove artifacts on the model update related to the presence of large amplitude double-scattered waves. Accounting more accurately for effects of the inverse Hessian operator may thus yield significant improvements. As a consequence, we present in this study an implementation of a matrix-free truncated Newton (TrN) method [5] for FWI. Instead of using an approximation of the inverse Hessian operator, this method partially solves the linear system (4) using a matrix-free conjugate gradient (CG) algorithm. This requires the capability of computing efficiently Hessian-vector products. This can be achieved using the second-order adjoint state formulas we present in the second section. In addition, an efficient adaptive stopping critertion for the CG iterations must be designed, in order to prevent from oversolving the equation (4), which would generate prohibitive computation costs. The Eisenstat stopping criterion [2] fulfills this requirement. Within this framework, numerical examples presented in the third section demonstrate that the computation cost of the l-BFGS method and the TrN method are comparable, and that the presence of large amplitude multi-scattered waves prevents the l-BFGS method from converging while the TrN method provides significantly more reliable results.

2. Method

2.1. Efficient computation of Hessian-vector products

In the following, we consider that $N_s = 1$ and we drop index s. Formulas for $N_s > 1$ are obtained straightforwardly by summation. We define the function $g_v(p)$ such that

$$g_v(p) = (\nabla f(p), v) = \mathcal{R}\left(J^\dagger R^\dagger (Ru(p) - d), v\right), \tag{5}$$

where $u(p)$ is the solution of (1), $J(p) = \partial_p u(p)$, $\mathcal{R}$ denotes the real part operator, and $\dagger$ the adjoint operator. We have $\nabla g_v(p) = \nabla^2 f(p)v$. The Lagrangian function associated with the functional $g_v(p)$ is

$$L_v(p,u,\alpha,\lambda,\mu) = \mathcal{R}\left(R^\dagger(Ru-d),\alpha\right) + \mathcal{R}\left(S(p)u-\varphi,\mu\right) + \mathcal{R}\left(S(p)\alpha + \sum_{j=1}^{m} v_j \partial_{p_j} S(p)u, \lambda\right), \tag{6}$$

where the first term is related to the function $g_v(p)$, the second term to the wave equation (1) and the third term to the first derivatives of (1). For $\widetilde{u}$ and $\widetilde{\alpha}$ such that

$$S(p)\widetilde{u} = \varphi, \quad S(p)\widetilde{\alpha} = -\sum_{j=1}^{m} v_j \partial_{p_j} S(p)\widetilde{u}, \tag{7}$$

we have

$$\nabla g_v(p) = \partial_p L_v(p,\widetilde{u},\widetilde{\alpha},\lambda,\mu) + \partial_u L_v(p,\widetilde{u},\widetilde{\alpha},\lambda,\mu)\,\partial_p\widetilde{u}(p) + \partial_\alpha L_v(p,\widetilde{u},\widetilde{\alpha},\lambda,\mu)\,\partial_p\widetilde{\alpha}(p). \tag{8}$$

We define $\widetilde{\lambda}$ and $\widetilde{\mu}$ such that

$$\partial_u L_v\left(p,\widetilde{u},\widetilde{\alpha},\widetilde{\lambda},\widetilde{\mu}\right) = 0, \quad \partial_\alpha L_v\left(p,\widetilde{u},\widetilde{\alpha},\widetilde{\lambda},\widetilde{\mu}\right) = 0. \tag{9}$$

We have

$$S(p)^\dagger\widetilde{\mu} = -R^\dagger R\widetilde{\alpha} - \sum_{j=1}^{m} v_j \left(\partial_{p_j} S(p)\right)^\dagger \widetilde{\lambda}, \quad S(p)^\dagger\widetilde{\lambda} = -R^\dagger(R\widetilde{u}-d), \tag{10}$$

and

$$(\nabla^2 f(p)v)_i = \mathcal{R}\left(\left((\partial_{p_i} S(p))\,\widetilde{u},\widetilde{\mu}\right) + \left((\partial_{p_i} S(p))\,\widetilde{\alpha},\widetilde{\lambda}\right) + \sum_{j=1}^{m} v_j \left(\left(\partial_{p_j}\partial_{p_i} S(p)\right)\widetilde{u},\widetilde{\lambda}\right)\right), i = 1,\dots,m. \tag{11}$$

Since computation of ∇f through the adjoint state method already requires the computation of $\widetilde{\lambda}$ and $\widetilde{u}$ [6], the computation of one matrix vector product $\nabla^2 f(p)v$ requires to solve one additional forward problem for $\widetilde{\alpha}$ and one additional adjoint problem for $\widetilde{\mu}$, as reported in [3, 4]. For $N_s > 1$, the overall computation cost is multiplied by N_s.

2.2. CG stopping criterion

An efficient implementation of the truncated Newton method requires the definition of an adaptive stopping criterion for the CG criterion. Newton methods are based on the sequential minimization of local quadratic approximations

$$q_k(\Delta p_k) = f(p_k) + (\nabla f(p_k),\Delta p_k) + (\nabla^2 f(p_k)\Delta p_k,\Delta p_k). \tag{12}$$

The accuracy required to solve the system (4) should reflect the accuracy of these local approximations to prevent from oversolving. This is achieved using the Eisenstat stopping criterion [2]:

$$\|H(p_k)\Delta p_k + \nabla f(p_k)\| \leq \eta_k \|\nabla f(p_k)\|, \tag{13}$$

where the forcing term η_k measures the distance between the misfit gradient and its first order development

$$\eta_k = \frac{\|\nabla f(p_k) - \nabla f(p_{k-1}) - H(p_{k-1})\Delta p_{k-1}\|}{\|\nabla f(p_{k-1})\|}. \tag{14}$$

Moreover, far from the solution, the Hessian operator should be indefinite. An additional stopping criterion is thus introduced: as soon as a negative curvature direction is computed during the resolution of the system (4), the CG iterations are stopped.

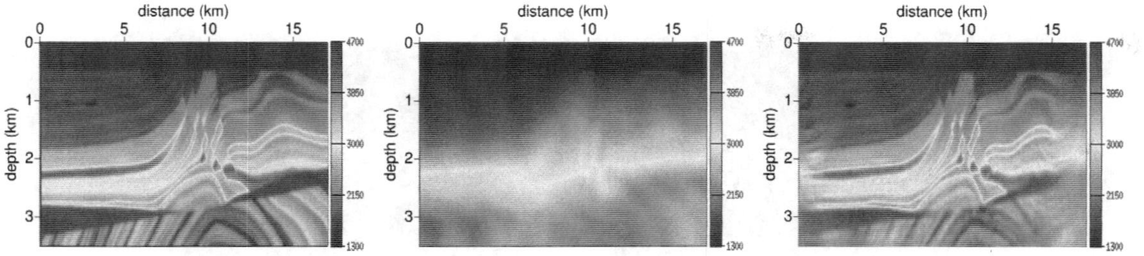

Figure 1. l-BFGS inversion results. Exact model (left), initial model (middle), estimated model (right).

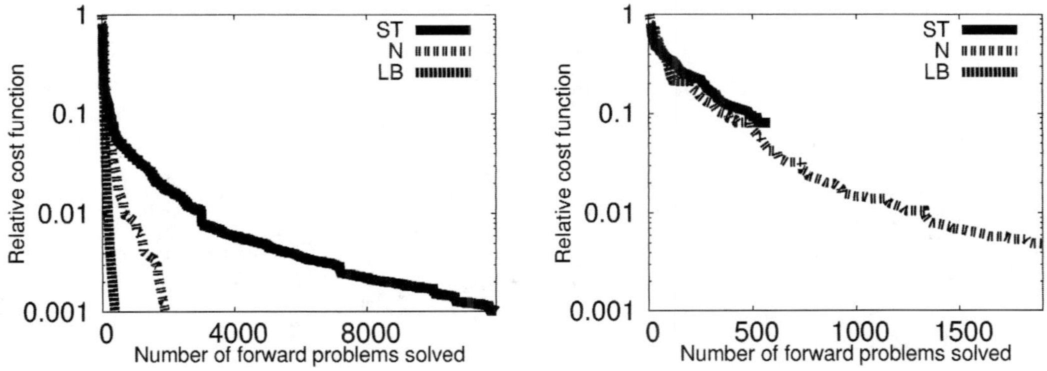

Figure 2. Convergence curves for the Marmousi II test case (left), for the near-surface imaging test case (right). N: truncated Newton method, LB: l-BFGS, ST: steepest-descent

3. Numerical results

3.1. The Marmousi II test case

The numerical results presented in this study are obtained in the 2D acoustic frequency domain FWI context. An estimation of the pressure wave velocity is computed. We first consider the Marmousi II benchmark test case. The Marmousi II model is 16 km wide and 3.5 km deep. The discretization step is set to 25 m. We use 144 sources and 660 receivers located near the surface to generate synthetic data. Four datasets corresponding to the frequencies $3, 5, 8, 12$ Hz are simultaneously inverted. The efficiencies of the steepest-descent, the l-BFGS algorithm and the TrN algorithm are compared. The three methods are implemented with the same linesearch globalization method. The iterations are stopped when the relative cost function $f(p)/f(p_0)$ reaches 10^{-3}. The estimated models obtained using the three different methods are very similar. The one obtained with the l-BFGS method is presented in figure 1.

The convergence curves (fig 2) are plotted as a function of the total number of the forward-problem resolutions. As expected, the steepest-descent algorithm converges very slowly, while the l-BFGS and the TrN method provide faster convergence. Note that the l-BFGS method is the fastest.

3.2. A near-surface imaging test case

We consider next a near-surface seismic imaging problem. Detecting and correctly imaging two concrete structures buried in the subsurface at few meters depth (fig. 4) is a challenge. The

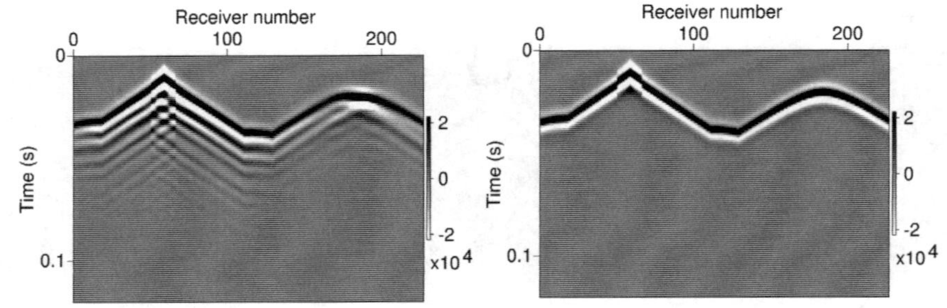

Figure 3. Dataset associated with the exact model (left), data associated with the initial homogeneous domain (right)

depth of investigation is limited to 3 m, the width of the exact model is 15 m. The discretization step is set to 0.15 m. We use a full acquisition system: four lines of sources/receivers are located on each side of the domain. We compute 9 datasets, from 100 Hz to 300 Hz each 25 Hz.

The very high velocity contrast between the background (300 $m.s^{-1}$) and the concrete foundations (4000 $m.s^{-1}$) generates energetic reflections. In addition, the close distance between the two structures is responsible for important multiple scattering. This is illustrated in figure 3 where two datasets in the time domain are presented[1], computed using the exact model and the homogeneous background model. The background model correctly predicts only the first-arrival waves. The signal below the first arrival on the left figure, corresponding to the multi-scattered waves, is not predicted. Starting from this background model, we invert simultaneously the 9 datasets, using the l-BFGS algorithm, and the TrN method. The convergence curves and the corresponding results are presented in figures 2 and 4 (the steepest-descent result is not presented since it is very similar to the result provided by the l-BFGS method). While the

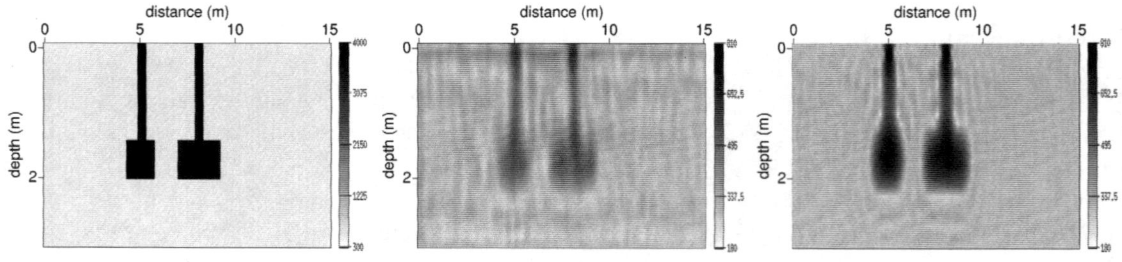

Figure 4. Pressure wave velocities. Exact model (left), l-BFGS estimation (middle), TrN estimation (right).

steepest-descent and the l-BFGS method stop after few iterations, trapped in a local minimum, the TrN method further reduces the misfit function. Why in this case the TrN method performs better than the l-BFGS method? Consider the Hessian operator expression:

$$\nabla^2 f(p) = \sum_{s=1}^{N_s} J_s^\dagger(p) R_s^\dagger R_s J_s(p) + \sum_{s=1}^{N_s} R_s^\dagger \left(R_s u_s(p) - d_s\right) \partial_{pp} u_s(p). \quad (15)$$

[1] These two datasets are generated using a Ricker source located at the surface between the two concrete structures at $x = 6.5$ m, centered on the frequency 200 Hz.

The first term of the Hessian operator is positive definite by construction, while the second one is indefinite, related to the double-scattered wavefield $\partial_{pp}u_s(p)$ and the residuals $R_s u_s(p) - d_s$. When single scattered waves dominate the data, gradient-based methods converge quickly from the first iterations. Therefore, the residuals decrease, and the Hessian operator tends to become positive definite. The l-BFGS methods builds a symmetric definite positive approximation of the inverse Hessian along the minimization process, and therefore improves the convergence speed of standard gradient methods. Since the Marmousi II model does not present high velocity contrasts, the amplitude of the multiscattered wavefield for this test case is weak, and the l-BFGS method is efficient.

When multiscattered waves cannot be neglected, as in our second test case, gradient-based methods face difficulties to fit the data, since the gradient direction only accounts for single scattered waves. Therefore, for the first iterations, the l-BFGS estimation only slowly decreases the residuals as the inverse Hessian approximation is close from the identity. As a consequence, the true Hessian operator stays indefinite while the l-BFGFS method build a definite positive approximation of its inverse. This explains why the l-BFGS method fails in this case. Conversely, the TrN method better accounts for the Hessian operator and is able to converge.

4. Conclusion and perspectives

Second-order adjoint formulas and the Eisenstat stopping criterion [2] yield an efficient implementation of the TrN method. Within this framework, the comparison of the TrN method with the l-BFGS method demonstrates the interest of using the TrN method when the seismic data is dominated by multi-scattered waves. In this case, the second-order information embedded in the Hessian operator, which is better accounted for by the TrN method, is crucial. The authors now look forward for the application of the TrN method to more general context, such as multi-parameter inversion and elastic FWI. The possibility of computing the inverse Hessian locally is also investigated in order to compute the posterior covariance matrix and provides an uncertainty estimation of the parameter reconstruction. Finally, the use of matrix-free preconditionners in order to speed-up the convergence of the CG algorithm will also be investigated.

Acknowledgments

This research is funded by the SEISCOPE consortium sponsored by BP, CGG-VERITAS, ENI, EXXON MOBIL, PETROBRAS, SAUDI ARAMCO, SHELL, STATOIL and TOTAL. The linear systems were solved with the MUMPS package. This work was performed by accessing to the high-performance computing facilities of CIMENT (Université de Grenoble, France) and to the HPC resources of GENCI-CINES under Grant 2011-046091.

References

[1] R. Byrd, P. Lu, and J. Nocedal, *A limited memory algorithm for bound constrained optimization*, SIAM Journal on Scientific and Statistical Computing, 16 (1995), pp. 1190–1208.

[2] S. C. Eisenstat and H. F. Walker, *Choosing the forcing terms in an inexact newton method*, SIAM Journal on Scientific Computing, 17 (1994), pp. 16–32.

[3] I. Epanomeritakis, V. Akçelik, O. Ghattas, and J. Bielak, *A Newton-CG method for large-scale three-dimensional elastic full waveform seismic inversion*, Inverse Problems, 24 (2008), pp. 1–26.

[4] A. Fichtner and J. Trampert, *Hessian kernels of seismic data functionals based upon adjoint techniques*, Geophysical Journal International, 185 (2011), pp. 775–798.

[5] S. G. Nash, *A survey of truncated Newton methods*, Journal of Computational and Applied Mathematics, 124 (2000), pp. 45–59.

[6] R. E. Plessix, *A review of the adjoint-state method for computing the gradient of a functional with geophysical applications*, Geophysical Journal International, 167 (2006), pp. 495–503.

[7] R. G. Pratt, C. Shin, and G. J. Hicks, *Gauss-Newton and full Newton methods in frequency-space seismic waveform inversion*, Geophysical Journal International, 133 (1998), pp. 341–362.

2nd Int. Workshop on New Computational Methods for Inverse Problems (NCMIP 2012) IOP Publishing
Journal of Physics: Conference Series **386** (2012) 012014 doi:10.1088/1742-6596/386/1/012014

Tomographic reconstruction of binary fields

Stéphane Roux, Hugo Leclerc and François Hild

Laboratoire de Mécanique et Technologie (LMT-Cachan)
ENS-Cachan/CNRS-UMR 8535/UPMC/PRES UniverSud Paris
61 avenue du Président Wilson, F-94235 Cachan Cedex, France.

E-mail: {stephane.roux;hugo.leclerc;francois.hild}@lmt.ens-cachan.fr

Abstract. A novel algorithm is proposed for reconstructing binary images from their projection along a set of different orientations. Based on a nonlinear transformation of the projection data, classical back-projection procedures can be used iteratively to converge to the sought image. A multiscale implementation allows for a faster convergence. The algorithm is tested on images up to 1 Mb definition, and an error free reconstruction is achieved with a very limited number of projection data, saving a factor of about 100 on the number of projections required for classical reconstruction algorithms.

1. Introduction

Binary reconstruction is a classical tomographic problem with the additional information that the reconstructed image is binary (*i.e.* 0-1 valued fields). The latter condition is a severe constraint that can help reducing drastically the number of projections needed for reconstruction *without* loss of information [1].

In the early 2000's, various algorithms have been proposed showing the feasibility of the reconstruction on modest definition images (typically 64^2 pixels). However, the proposed methods are generally based on sophisticated optimization strategies of multiple minima problems, *e.g.* simulated annealing, genetic programming [2]. Hence, the very slow convergence (typically hours) and poor success rate (few percent remaining errors) tended to confine such techniques to specific cases where projection data acquisition is limited.

Besides, in a seminal contribution, Candés *et al.* [3] have shown that images having a few number of gray levels could be reconstructed *exactly* from a very limited number of projections. The proposed strategy was to use a regularization based on the total variation of the image. Independently, Batenburg [4] proposed a specific algebraic algorithm to reconstruct binary images that turns out to be extremely efficient as compared to alternative published algorithms, allowing for larger images to be considered. In the line of the latter algorithm, a new approach is proposed herein to deal with large definition images requiring minutes on a standard PC, thereby allowing for a much larger class of applications. For example, one-megapixel images are reconstructed *error-free* from a few projections (the number depends on the "complexity" of the image pattern, but 5-20 is a typical order of magnitude) in a few minutes on a standard PC without sophisticated implementation.

2. Proposed algorithm

The binary image is denoted as $f(\boldsymbol{x})$, such that at a pixel located in $\boldsymbol{x}$, the gray level value f is either 0 or 1. A projection in the direction $\boldsymbol{n}_\theta$ is defined as the sum of $f(\boldsymbol{x})$ for all pixels $\boldsymbol{x}$ located along the same

Journal of Physics: Conference Series **386** (2012) 012014 doi:10.1088/1742-6596/386/1/012014

ray of direction $\boldsymbol{n}_\theta$. More precisely, introducing the unit vector $\boldsymbol{t}_\theta$ normal to $\boldsymbol{n}_\theta$, the projection $S(y,\theta)$ is written as

$$\begin{aligned} S(y,\theta) &= \int f(\boldsymbol{x})\delta(\boldsymbol{x}\cdot\boldsymbol{t}_\theta - y)\,\mathrm{d}\boldsymbol{x} \\ &\equiv \int P_\theta(y,\boldsymbol{x})f(\boldsymbol{x})\,\mathrm{d}\boldsymbol{x} \end{aligned} \tag{1}$$

where P_θ is the projection operator. For each direction θ, the detected signal is discretized with a bin size equal to the pixel size. In the following, for the sake of simplicity a pixel $\boldsymbol{x}$ is considered to project on exactly one detector site y (the closest to its projected position $\boldsymbol{x}\cdot\boldsymbol{t}_\theta$) so that the sparse matrix P_θ is itself binary-valued. Moreover, non-zero values of f are assumed to lie with a known domain, $\mathcal{D}$, the indicator function of which is characterized by its projection $S_0(y,\theta) = P_\theta(y,\boldsymbol{x})\mathbb{1}_{\mathcal{D}}(\boldsymbol{x})$. The problem consists in finding $f(\boldsymbol{x})$ from a collection of projection vectors $S(y,\theta)$, for a few evenly distributed θ angles.

2.1. Initialization

The algorithm consists in an initialization step and an iterative correction loop. The initialization step constructs the probabilities ϖ_i that a given pixel i be 1, as can be estimated from the different projections. Each projection θ_j passing through the site independently provides such a probability $p_j = S(y,\theta_j)/S_0(y,\theta_j)$. It can be shown that when two projections are available for the same site, the probability for the site to be valued 1 is

$$\varpi = \frac{p_1 p_2}{p_1 p_2 + (1-p_1)(1-p_2)} \tag{2}$$

These probabilities are very conveniently represented by their nonlinear transforms

$$\Phi(p) \equiv \log\left(\frac{p}{(1-p)}\right) \tag{3}$$

since the combination of probabilities, Eq. (2), is equivalent to

$$\Phi(\varpi) = \Phi(p_1) + \Phi(p_2) \tag{4}$$

This argument holds for an arbitrary number of projections, so that the probability ϖ_i is obtained from

$$\varpi = \Phi^{-1}\left(\sum_{projection\ i} \Phi(p_i)\right) \tag{5}$$

Thus the initial probability field is obtained from a mere back-projection algorithm of the transformed projections, $\Phi(p)$, and the final probability field is retrieved from the Φ^{-1} nonlinear transform.

The resulting probability field is then filtered by a convolution with a Gaussian filter with a short characteristic scale δ typically chosen to be of the order of a few pixels. The resulting field is thresholded setting large ($p > 0.99$) and low ($p < 0.01$) probabilities to 1 and 0, respectively. A second determination of the intermediate probabilities is made securing the previous thresholded pixels (namely, S and S_0 are corrected from the contributions of these known pixels, and only undetermined pixels are updated with a new estimate of ϖ).

2.2. Correction

The second step is a correction loop where the spirit consists in maximizing the similarity with the filtered p-field while satisfying the projection constraint independently for each projection direction. After the initialization step, it is not possible to simply threshold the probability field ϖ at 1/2 (or equivalently $\Phi(\varpi)$ at 0) to obtain f (since in particular the nonlinear transformation does not preserve the expectation value). The principle of the algorithm consists in correcting the p values along each ray (y,θ) so that a binarization of ϖ would match the known ray projection $S(y, \theta)$. This nonlinear problem reduces to translating uniformly all the $\Phi(\varpi)$ values along the ray so that the $S(y, \theta)$ largest ones become positive. After sweeping through all directions, the estimate $\tilde{f}$ of f is constructed from the sign of $\Phi(\varpi)$ ($\tilde{f} = 1$ for positive values of $\Phi(\varphi)$ and $\tilde{f} = 0$ for negative ones). As long as $\tilde{f}$ is not consistent with the projection information, this correction step is repeated, namely, ϖ is first evaluated from the convolution of the current estimate of $\tilde{f}$ by a Gaussian, and $\Phi(\varpi)$ are corrected as indicated above. As the number of iteration increases, the Gaussian used in the convolution is scaled down to a width of 1 pixel. Typically, a few tens of such steps are required to reach an error-free image. It is emphasized that the recourse to the Φ-transformed projection data (and probabilities) — rather than ϖ — is a crucial point for a fast convergence.

2.3. Multiscale strategy

This algorithm can be used in a multiscale strategy to tackle images with complex texture and large sizes, although simple images are more efficiently reconstructed at the finest level of description. The spirit of the multiscale approach is to coarsen the projections gathering two consecutive y's into one $S'(y, \theta) = (S(2y, \theta) + S(2y + 1, \theta))/4$, and averaging 2×2 pixels in the image f into "super-pixels" in a recursive fashion. Each level of this pyramidal construction is termed a "generation," and the finest one is labeled 0. Starting from the coarsest scale, a binary reconstruction is first performed as above presented, and the result used as a predetermination of the refined level.

3. Example

This algorithm was tested on a series of artificial domains of variable complexity. Unions of ellipses or polygons were used to generate microstructures as they were used as a benchmark proposed by Batenburg [4] and compared with alternative algorithms. In all those cases, the presented algorithm revealed to perform better in terms of time (and error when a fixed number of iterations was prescribed) as compared to all reported methods in Ref. [4].

However, the choice of the image microstructures is often quite simple as compared to what may be encountered in practical examples. For complex microstructures and large image sizes, it is essential to use this multiscale strategy. In order to test a realistic case, an example shown in Figure 1 and extracted from real tomographic data is used as an illustration. The full image as reconstructed from a Filtered Back Projection algorithm is first binarized and filtered using a phase-field approach favoring a two level image and taking into account a line tension. This example is demanding because it contains a large number of disconnected domains with small gaps between them, and the domains have a rather wide size distribution down to very small ones. Moreover, the domain size is quite large.

Table 1 gives the number of iterations and computation time to reconstruct the binary image shown in Figure 1. This image is identical (for all single pixel) to the binary image issued from the initial binarization although it is reconstructed from 17 projections only. Note that the original data shown in Fig. 1(left) required 1500 projections. Moreover the image size is 1025×1025 pixels, a size much larger than what is usually considered in the literature. Table 1 also reports the time needed to reach convergence (*i.e.* not a single pixel difference between reference and reconstruction) at each generation. The present code is a simple Matlab® program run on a standard single processor PC without much optimization.

The advantage of the multiscale procedure can be seen in the fact that the same algorithm does not converge when the number of coarse-graining step is 0 (*i.e.* no coarse-graining) or 1. Choosing a number of generations from 2 to 5 results in a total computation time of 259, 134, 134 and 133 s respectively. Although the benefit in time is negligible when increasing the number of generations from 3 to 5, robustness increases and hence a large number of scales is preferred. Robustness means here that the number of coarse-graining steps does not need any tuning depending on the image texture. A large number of generations is at worse useless, but never a penalty for time or performance.

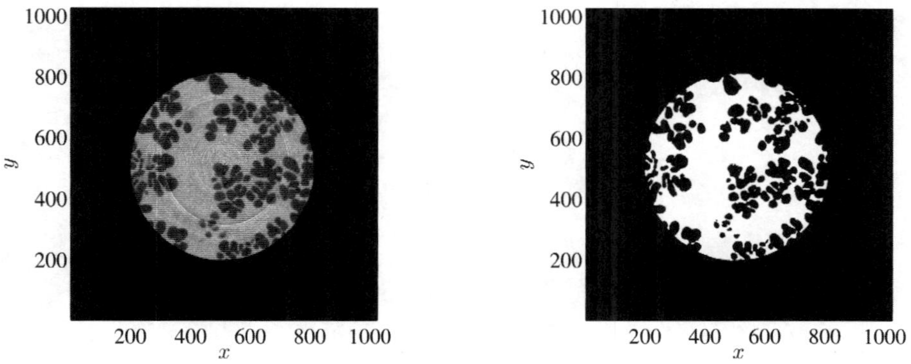

Figure 1. An example of a real tomographic image (Courtesy of E. Gouillart and C. Zang) and its binary version.

Table 1. Number of iterations and execution time at each level of the multiscale algorithm. At each level an error-free reconstruction is obtained.

Generation	Iteration	Time (s)
5	9	2.64
4	19	2.73
3	25	9.56
2	13	16.9
1	7	31.5
0	4	69.7
Total		133

4. Conclusion

A novel binary reconstruction algorithm has been presented. A direct algebraic approach, based on the probability field that pixels be 0/1-valued, is proposed. The major difference with earlier approaches is the recourse to a nonlinear transform of the probability field. Such a transform is used for the initialization step and for the correction one. This leads to a significant increase in the performances allowing for error-free high definition image reconstructions (*i.e.* larger then 1 Mb) in minutes without any fancy optimization of the implementation.

Acknowledgments

Communication of the raw tomographic data shown in Fig. 1 by E. Gouillart (CNRS/Saint-Gobain, Aubervilliers, France) and C. Zang (RWTH, Aachen, Germany) is gratefully acknowledged. This work was supported by the Agence Nationale de la Recherche, through "RUPXCUBE" and "EDDAM" projects (ANR-09-BLAN-0009-03 and ANR-11-BS09-027).

References

[1] A.C. Kak, M. Slaney, Algorithms for Reconstruction with Non-Diffracting Sources, in: Principles of computerized tomographic imaging, IEEP Press, New York (NY), USA, (1988).
[2] G.T. Herman, A. Kuba (Eds.), Advances in Discrete Tomography and Its Applications, Birkhaüser, (Boston, USA) (2007)
[3] E. Candès, J. Romberg, T. Tao, IEEE Transactions on information theory **52**, (2006) 489-509
[4] K.J. Batenburg, J. Math. Imaging Vis. **27**, (2007) 175-191

AUTHOR INDEX

Institute of Physics
Dirac House, Temple Back
Bristol BS1 6BE UK

CURRAN ASSOCIATES INC
proceedings
.com